Idea Makers

Personal Perspectives on the Lives & Ideas of Some Notable People

Stephen Wolfram

Idea Makers

Personal Perspectives on the Lives & Ideas of Some Notable People

Idea Makers: Personal Perspectives on the Lives & Ideas of Some Notable People

Wolfram Media, Inc. | wolfram-media.com

ISBN 978-1-57955-003-5 (hardback)
ISBN 978-1-57955-005-9 (ebook)

Biography / Science

Library of Congress Cataloging-in-Publication Data

Wolfram, Stephen, author.
Idea makers : personal perspectives on the lives & ideas of some notable people by Stephen Wolfram. First edition. | Champaign : Wolfram Media, Inc., [2016] LCCN 2016025380 (print) | LCCN 2016026486 (ebook) ISBN 9781579550035 (hardcover : acid-free paper) ISBN 9781579550059 (ebook) | ISBN 9781579550110 (ePub) ISBN 9781579550059 (kindle) LCSH: Scientists—Biography. | Science—History. LCC Q141 .W678562 2016 (print) | LCC Q141 (ebook) DDC 509.2/2—dc23 LC record available at https://lccn.loc.gov/2016025380

Stephen Wolfram. Idea Makers: Personal Perspectives on the Lives & Ideas of Some Notable People (Kindle Locations 10-21). Wolfram Media, Inc. | wolfram-media.com. Kindle Edition.

Sources for photos and archival materials that are not from the author's collection or in the public domain:

pp. 45, 66, 67: The Carl H. Pforzheimer Collection of Shelley and His Circle, The New York Public Library; pp. 46, 55, 56, 63, 65, 67, 79: Additional Manuscripts Collection, Charles Babbage Papers, British Library; p. 49: National Portrait Gallery; pp. 50, 60: Museum of the History of Science, Oxford; pp. 52, 53, 85, 86: Science Museum/Science & Society Picture Library; p. 54: The Power House Museum, Sydney; p. 68: Lord Lytton, The Bodleian Library; pp. 99–120: Leibniz-Archiv/Leibniz Research Center Hannover, State Library of Lower Saxony; p. 138: Alisa Bokulich; p. 167–170, 178: Cambridge University Library; pp. 192, 193: Trinity College Library; p. 194: Tata Institute of Fundamental Research

---

Printed by Friesens, Manitoba, Canada. ♾ Acid-free paper. First edition. Second printing.

Table of Contents

Preface

I've spent most of my life working hard to build the future with science and technology. But two of my other great interests are history and people. This book is a collection of essays I've written that indulge those interests. All of them are in the form of personal perspectives on people—describing from my point of view the stories of their lives and of the ideas they created.

I've written different essays for different reasons: sometimes to commemorate a historical anniversary, sometimes because of a current event, and sometimes—unfortunately—because someone just died. The people I've written about span three centuries in time—and range from the very famous to the little-known. All of them had interests that intersect in some way or another with my own. But it's ended up a rather eclectic list—that's given me the opportunity to explore a wide range of very different lives and ideas.

When I was younger, I really didn't pay much attention to history. But as the decades have gone by, and I've seen so many different things develop, I've become progressively more interested in history—and in what it can teach us about the pattern of how things work. And I've learned that decoding the actual facts and path of history—like so many other areas—is a fascinating intellectual process.

There's a stereotype that someone focused on science and technology won't be interested in people. But that's not me. I've always been interested in people. I've been fortunate over the course of my life to get to know a very large and diverse set of them. And as I've grown my company over the past three decades I've had the pleasure of working with many wonderful individuals. I always like to give help and advice. But I'm also fascinated just to watch the trajectories of people's lives—and to see how people end up doing the things they do.

It's been great to personally witness so many life trajectories over the past half century. And in this book I've written about a few of them.

But I've also been interested to learn about the life trajectories of those from the more distant past. Usually I know quite a lot about the end of the story: the legacy of their work and ideas. But I find it fascinating to see how these things came to be—and how the paths of people's lives led to what they did.

Part of my interest is purely intellectual. But part of it is more practical—and more selfish. What can I learn from historical examples about how things I'm involved in now will work out? How can I use people from the past as models for people I know now? What can I learn for my own life from what these people did in their lives?

To be clear: this book is not a systematic analysis of great thinkers and creators through history. It is an eclectic collection of essays about particular people who for one reason or another were topical for me to write about. I've tried to give both a sketch of each person's life in its historical context and a description of their ideas—and then I've tried to relate those ideas to my own ideas, and to the latest science and technology.

In the process of writing these essays I've ended up doing a considerable amount of original research. When the essays are about people I've personally known, I've been able to draw on interactions I had with them, as well as on material I've personally archived. For other people, I've tried when it's possible to seek out individuals who knew them—and in all cases I've worked hard to find original documents and other primary sources. Many people and institutions have been very forthcoming with their help—and also it's been immensely helpful that in modern times so many historical documents have been scanned and put on the web.

But with all of this, I'm still constantly struck by how hard it is to do history. So often there's been some story or analysis that people repeat all the time. But somehow something about it hasn't quite rung true with me. So I've gone digging to try to find out the real story. Occasionally one just can't tell what it was. But at least for the people I've written about in this book, there are usually enough records and documents—or actual people to talk to—that one can eventually figure it out.

My strategy is to keep on digging and getting information until things make sense to me, based on my knowledge of people and situations that are somehow similar to what I'm studying. It's certainly helped that in my own life I've seen all sorts of ideas and other things develop over the course of years—which has given me some intuition about how such things work. And one of the important lessons of this is that however brilliant one may be, every idea is the result of some progression or path—often hard-won. If there seems to be a jump in the story—a missing link—then that's just because one hasn't figured it out. And I always try to go on until there aren't mysteries anymore, and everything that happened makes sense in the context of my own experiences.

So having traced the lives of quite a few notable people, what have I learned? Perhaps the clearest lesson is that serious ideas that people have are always deeply entwined with the trajectories of their lives. That is not to say that people always live the paradigms they create—in fact, often, almost paradoxically, they don't. But ideas arise out of the context of people's lives. Indeed, more often than not, it's a very practical situation that someone finds themselves in that leads them to create some strong, new, abstract idea.

When history is written, all that's usually said is that so-and-so came up with such-and-such an idea. But there's always more to it: there's always a human story behind it. Sometimes that story helps illuminate the abstract idea. But more often, it instead gives us insight about how to turn some human situation or practical issue into something intellectual—and perhaps something that will live on, abstractly, long after the person who created it is gone.

This book is the first time I've systematically collected what I've written about people. I've written more generally about history in a few other places—for example in the hundred pages or so of detailed historical notes at the back of my 2002 book *A New Kind of Science*. I happened to start my career young, so my early colleagues were often much older than me—making it demographically likely that there may be many obituaries for me to write. But somehow I find it cathartic to reflect on how a particular life added stones—large or small—to the great tower that represents our civilization and its achievements.

I wish I could have personally known all the people I write about in this book. But for those who died long ago it feels like a good second best to read so many documents they wrote—and somehow to get in and understand their lives. My greatest personal passion remains trying to build the future. But I hope that through understanding the past I may be able to do it a little better—and perhaps help build it on a more informed and solid basis. For now, though, I'm just happy to have been able to spend a little time on some remarkable people and their remarkable lives—and I hope that we'll all be able to learn something from them.

Stephen Wolfram

For external links and references, see stephenwolfram.com, or the ebook version of this book.

Richard Feynman

April 20, 2005

I first met Richard Feynman when I was 18, and he was 60. And over the course of ten years, I think I got to know him fairly well. First when I was in the physics group at Caltech. And then later when we both consulted for a once-thriving Boston company called Thinking Machines Corporation.

I actually don't think I've ever talked about Feynman in public before. And there's really so much to say, I'm not sure where to start.

But if there's one moment that summarizes Richard Feynman and my relationship with him, perhaps it's this. It was probably 1982. I'd been at Feynman's house, and our conversation had turned to some kind of unpleasant situation that was going on. I was about to leave. And Feynman stopped me and said, "You know, you and I are very lucky. Because whatever else is going on, we've always got our physics."

Feynman loved doing physics. I think what he loved most was the process of it. Of calculating. Of figuring things out. It didn't seem to matter to him so much if what came out was big and important. Or esoteric and weird. What mattered to him was the process of finding it. And he was often quite competitive about it.

Some scientists (myself probably included) are driven by the ambition to build grand intellectual edifices. I think Feynman—at least in the years I knew him—was much more driven by the pure pleasure of actually doing the science. He seemed to like best to spend his time figuring things out, and calculating. And he was a great calculator. All around perhaps the best human calculator there's ever been.

A talk given on the occasion of the publication of Feynman's collected letters.

Here's a page from my files: quintessential Feynman. Calculating a Feynman diagram:

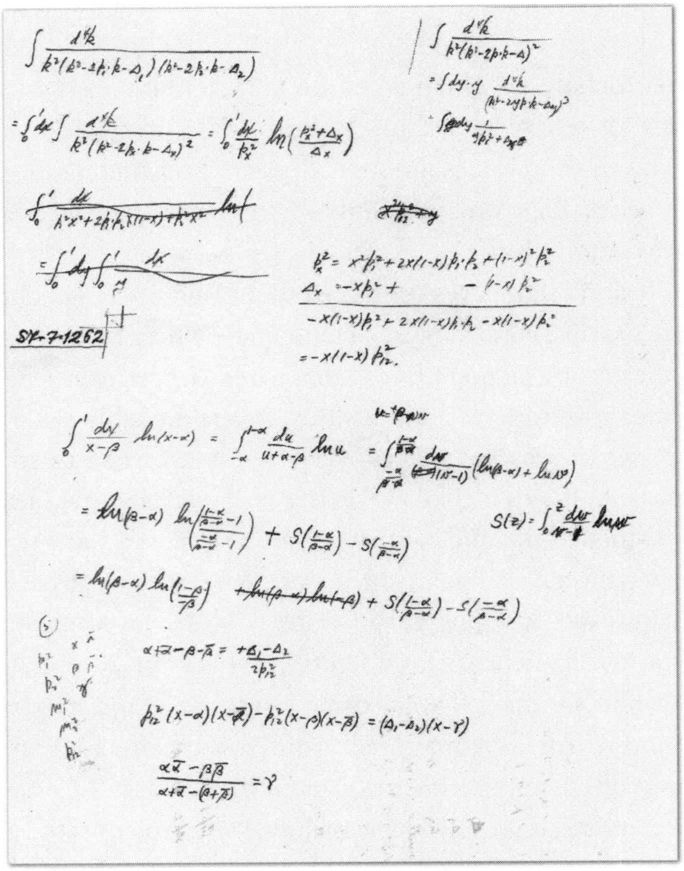

It's kind of interesting to look at. His style was always very much the same. He always just used regular calculus and things. Essentially nineteenth-century mathematics. He never trusted much else. But wherever one could go with that, Feynman could go. Like no one else.

I always found it incredible. He would start with some problem, and fill up pages with calculations. And at the end of it, he would actually get the right answer! But he usually wasn't satisfied with that. Once he'd gotten the answer, he'd go back and try to figure out why it was obvious. And often he'd come up with one of those classic Feynman straightforward-sounding explanations. And he'd never tell people about all the calculations behind it. Sometimes it was kind of a game for him: having people be flabbergasted by his seemingly instant physical intuition, not knowing that really it was based on some long, hard calculation he'd done.

He always had a fantastic formal intuition about the innards of his calculations. Knowing what kind of result some integral should have, whether some special case should matter, and so on. And he was always trying to sharpen his intuition.

You know, I remember a time—it must have been the summer of 1985—when I'd just discovered a thing called rule 30. That's probably my own all-time favorite scientific discovery. And that's what launched a lot of the whole new kind of science that I've spent 20 years building (and wrote about in my book *A New Kind of Science*).

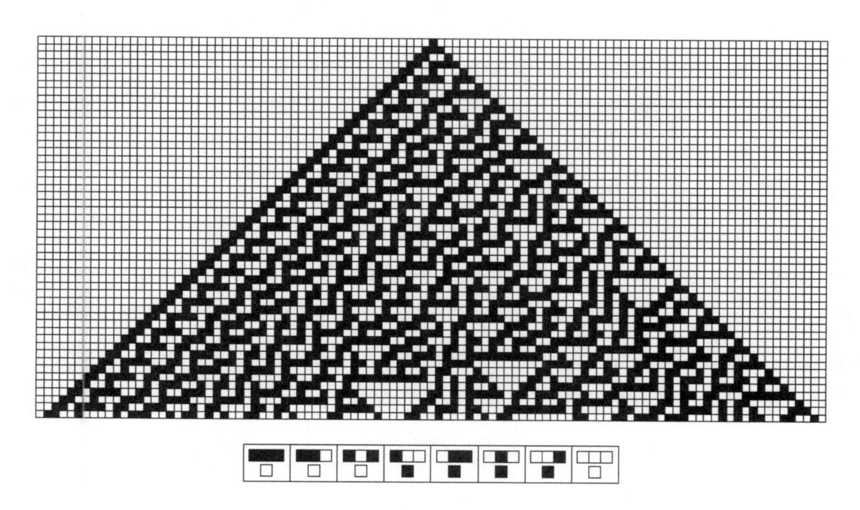

Well, Feynman and I were both visiting Boston, and we'd spent much of an afternoon talking about rule 30. About how it manages to go from that little black square at the top to make all this complicated stuff. And about what that means for physics and so on.

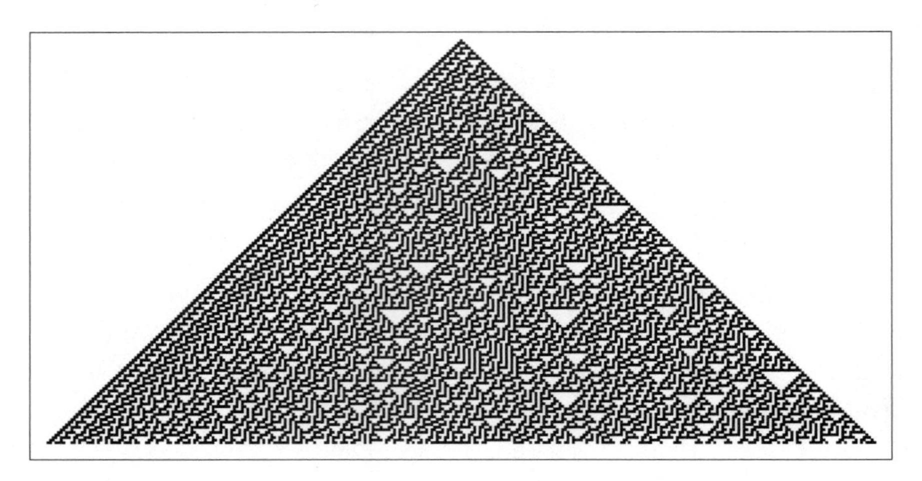

Well, we'd just been crawling around the floor—with help from some other people—trying to use meter rules to measure some feature of a giant printout of it. And Feynman took me aside, rather conspiratorially, and said, "Look, I just want to ask you one thing: how did you know rule 30 would do all this crazy stuff?" "You know me," I said. "I didn't. I just had a computer try all the possible rules. And I found it." "Ah," he said, "now I feel much better. I was worried you had some way to figure it out."

Feynman and I talked a bunch more about rule 30. He really wanted to get an intuition for how it worked. He tried bashing it with all his usual tools. Like he tried to work out what the slope of the line between order and chaos is. And he calculated. Using all his usual calculus and so on. He and his son Carl even spent a bunch of time trying to crack rule 30 using a computer.

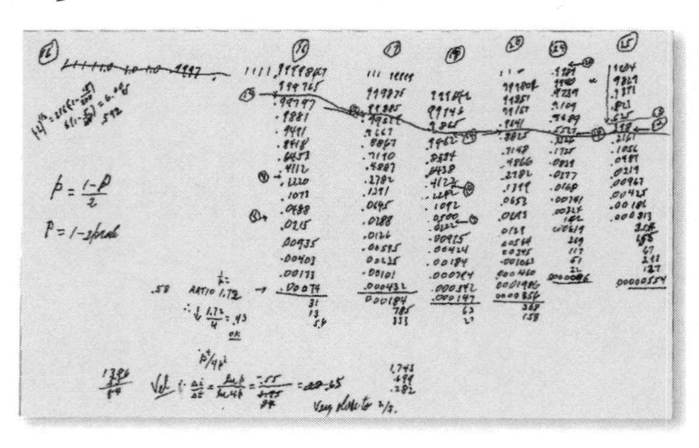

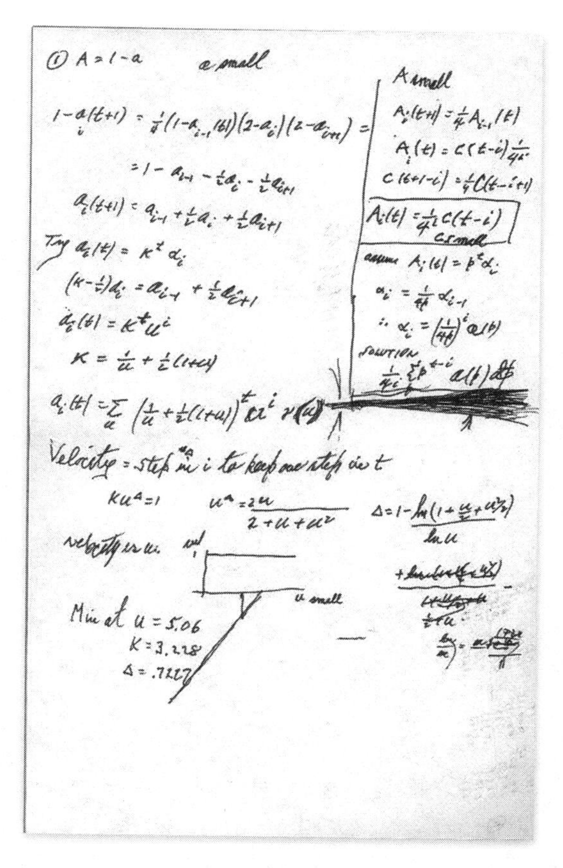

And one day he calls me and says, "OK, Wolfram, I can't crack it. I think you're on to something." Which was very encouraging.

Feynman and I tried to work together on a bunch of things over the years. On quantum computers before anyone had ever heard of those. On trying to make a chip that would generate perfect physical randomness—or eventually showing that that wasn't possible. On whether all the computation needed to evaluate Feynman diagrams really was necessary. On whether it was a coincidence or not that there's an e^{-Ht} in statistical mechanics and an e^{iHt} in quantum mechanics. On what the simplest essential phenomenon of quantum mechanics really is.

I remember often when we were both consulting for Thinking Machines in Boston, Feynman would say, "Let's hide away and do some physics." This was a typical scenario. Yes, I think we thought nobody was noticing that we were off at the back of a press conference about a new computer system talking about the nonlinear sigma model. Typically, Feynman would do some calculation. With me continually protesting that we should just go and use a computer. Eventually I'd do that. Then I'd get some results. And he'd get some results. And then we'd have an argument about whose intuition about the results was better.

I should say, by the way, that it wasn't that Feynman didn't like computers. He even had gone to some trouble to get an early Commodore PET personal computer, and enjoyed doing things with it. And in 1979, when I started working on the forerunner of what would become Mathematica, he was very interested. We talked a lot about how it should work. He was keen to explain his methodologies for solving problems: for doing integrals, for notation, for organizing his work. I even managed to get him a little interested in the problem of language design. Though I don't think there's anything directly from Feynman that has survived in Mathematica. But his favorite integrals we can certainly do.

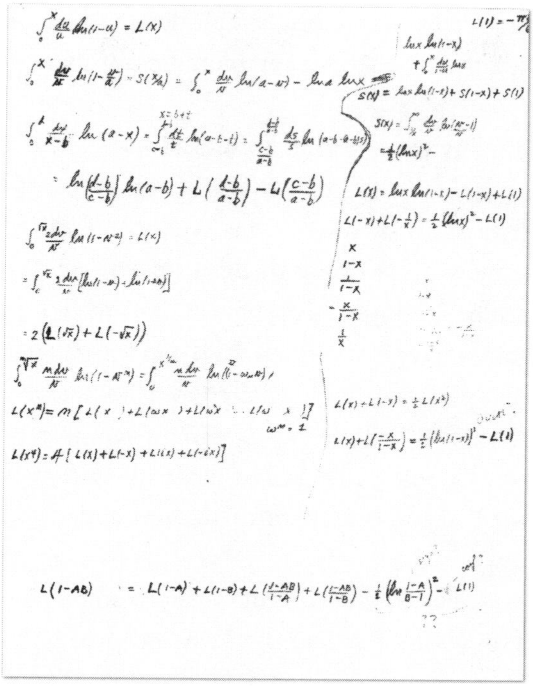

In[1]:= $\int \frac{\text{Log}[a - x]}{x - b}\, dx$

Out[1]= $\text{Li}_2\left(\frac{a - x}{a - b}\right) + \log(a - x)\log\left(1 - \frac{a - x}{a - b}\right)$

You know, it was sometimes a bit of a liability having Feynman involved. Like when I was working on SMP—the forerunner of Mathematica—I organized some seminars by people who'd worked on other systems. And Feynman used to come. And one day a chap from a well-known computer science department came to speak. I think he was a little tired, and he ended up giving what was admittedly not a good talk. And it degenerated at some point into essentially telling puns about the name of the system they'd built. Well, Feynman got more and more annoyed. And eventually stood up and gave a whole speech about how "If this is what computer science is about, it's all nonsense..." I think the chap who gave the talk thought I'd put Feynman up to this. And has hated me for 25 years...

You know, in many ways, Feynman was a loner. Other than for social reasons, he really didn't like to work with other people. And he was mostly interested in his own work. He didn't read or listen too much; he wanted the pleasure of doing things himself. He did used to come to physics seminars, though. Although he had rather a habit of using them as problem-solving exercises. And he wasn't always incredibly sensitive to the speakers. In fact, there was a period of time when I organized the theoretical physics seminars at Caltech. And he often egged me on to compete to find fatal flaws in what the speakers were saying. Which led to some very unfortunate incidents. But also led to some interesting science.

One thing about Feynman is that he went to some trouble to arrange his life so that he wasn't particularly busy—and so he could just work on what he felt like. Usually he had a good supply of problems. Though sometimes his long-time assistant would say: "You should go and talk to him. Or he's going to start working on trying to decode Mayan hieroglyphs again." He always cultivated an air of irresponsibility. Though I would say more towards institutions than people.

And I was certainly very grateful that he spent considerable time trying to give me advice—even if I was not always great at taking it. One of the things he often said was that "peace of mind is the most important prerequisite for creative work." And he thought one should do everything one could to achieve that. And he thought that meant, among other things, that one should always stay away from anything worldly, like management.

Feynman himself, of course, spent his life in academia—though I think he found most academics rather dull. And I don't think he liked their standard view of the outside world very much. And he himself often preferred more unusual folk.

Quite often he'd introduce me to the odd characters who'd visit him. I remember once we ended up having dinner with the rather charismatic founder of a semi-cult called EST. It was a curious dinner. And afterwards, Feynman and I talked for hours about leadership. About leaders like Robert Oppenheimer. And Brigham Young. He was fascinated—and mystified—by what it is that lets great leaders lead people to do incredible things. He wanted to get an intuition for that.

You know, it's funny. For all Feynman's independence, he was surprisingly diligent. I remember once he was preparing some fairly minor conference talk. He was quite concerned about it. I said, "You're a great speaker; what are you worrying about?" He said, "Yes, everyone thinks I'm a great speaker. So that means they expect more from me." And in fact, sometimes it was those throwaway conference talks that have ended up being some of Feynman's most popular pieces. On nanotechnology. Or foundations of quantum theory. Or other things.

You know, Feynman spent most of his life working on prominent current problems in physics. But he was a confident problem solver. And occasionally he would venture outside, bringing his "one can solve any problem just by thinking about it" attitude with him. It did have some limits, though. I think he never really believed it applied to human affairs, for example. Like when we were both consulting for Thinking Machines in Boston, I would always be jumping up and down about how if the management of the company didn't do this or that, they would fail. He would just say, "Why don't you let these people run their company; we can't figure out this kind of stuff." Sadly, the company did in the end fail. But that's another story.

Kurt Gödel

May 1, 2006

When Kurt Gödel was born—one hundred years ago today—the field of mathematics seemed almost complete. Two millennia of development had just been codified into a few axioms, from which it seemed one should be able almost mechanically to prove or disprove anything in mathematics—and, perhaps with some extension, in physics too.

Twenty-five years later things were proceeding apace, when at the end of a small academic conference, a quiet but ambitious fresh PhD involved with the Vienna Circle ventured that he had proved a theorem that this whole program must ultimately fail.

In the seventy-five years since then, what became known as Gödel's theorem has been ascribed almost mystical significance, sowed the seeds for the computer revolution, and meanwhile been practically ignored by working mathematicians—and viewed as irrelevant for broader science.

The ideas behind Gödel's theorem have, however, yet to run their course. And in fact I believe that today we are poised for a dramatic shift in science and technology for which its principles will be remarkably central.

Gödel's original work was quite abstruse. He took the axioms of logic and arithmetic, and asked a seemingly paradoxical question: can one prove the statement "this statement is unprovable"?

One might not think that mathematical axioms alone would have anything to say about this. But Gödel demonstrated that in fact his statement could be encoded purely as a statement about numbers.

Yet the statement says that it is unprovable. So here, then, is a statement within mathematics that is unprovable by mathematics: an "undecidable statement". And its existence immediately shows that there is a certain incompleteness to mathematics: there are mathematical statements that mathematical methods cannot reach.

It could have been that these ideas would rest here. But from within the technicalities of Gödel's proof there emerged something of incredible practical importance. For Gödel's seemingly bizarre technique of encoding statements in terms of numbers was a critical step towards the idea of universal computation—which implied the possibility of software, and launched the whole computer revolution.

Thinking in terms of computers gives us a modern way to understand what Gödel did: although he himself in effect only wanted to talk about one computation, he proved that logic and arithmetic are actually sufficient to build a universal computer, which can be programmed to carry out any possible computation.

Not all areas of mathematics work this way. Elementary geometry and elementary algebra, for example, have no universal computation, and no analog of Gödel's theorem—and we even now have practical software that can prove any statement about them.

But universal computation—when it is present—has many deep consequences.

The exact sciences have always been dominated by what I call computational reducibility: the idea of finding quick ways to compute what systems will do. Newton showed how to find out where an (idealized) Earth will be in a million years—we just have to evaluate a formula, we do not have to trace a million orbits.

But if we study a system that is capable of universal computation we can no longer expect to "outcompute" it like this; instead, to find out what it will do may take us irreducible computational work.

And this is why it can be so difficult to predict what computers will do—or to prove that software has no bugs. It is also at the heart of why mathematics can be difficult: it can take an irreducible amount of computational work to establish a given mathematical result.

And it is what leads to undecidability—for if we want to know, say, whether any number of any size has a certain property, computational irreducibility may tell us that without checking infinitely many cases we may not be able to decide for sure.

Working mathematicians, though, have never worried much about undecidability. For certainly Gödel's original statement is remote, being astronomically long when translated into mathematical form. And the few alternatives constructed over the years have seemed almost as irrelevant in practice.

But my own work with computer experiments suggests that in fact undecidability is much closer at hand. And indeed I suspect that quite a few of the famous unsolved problems in mathematics today will turn out to be undecidable within the usual axioms.

The reason undecidability has not been more obvious is just that mathematicians—despite their reputation for abstract generality—like most scientists, tend to concentrate on questions that their methods succeed with.

Back in 1931, Gödel and his contemporaries were not even sure whether Gödel's theorem was something general, or just a quirk of their formalism for logic and arithmetic. But a few years later, when Turing machines and other models for computers showed the same phenomenon, it began to seem more general.

Still, Gödel wondered whether there would be an analog of his theorem for human minds, or for physics. We still do not know the complete answer, though I certainly expect that both minds and physics are in principle just like universal computers—with Gödel-like theorems.

One of the great surprises of my own work has been just how easy it is to find universal computation. If one systematically explores the abstract universe of possible computational systems one does not have to go far. One needs nothing like the billion transistors of a modern electronic computer, or even the elaborate axioms of logic and arithmetic. Simple rules that can be stated in a short sentence—or summarized in a three-digit number—are enough.

And it is almost inevitable that such rules are common in nature—bringing with them undecidability. Is a solar system ultimately stable? Can a biochemical process ever go out of control? Can a set of laws have a devastating consequence? We can now expect general versions of such questions to be undecidable.

This might have pleased Gödel—who once said he had found a bug in the US Constitution, who gave his friend Einstein a paradoxical model of the universe for his birthday—and who told a physicist I knew that for theoretical reasons he "did not believe in natural science".

Even in the field of mathematics, Gödel—like his results—was always treated as somewhat alien to the mainstream. He continued for decades to supply central ideas to mathematical logic, even as "the greatest logician since Aristotle" (as John von Neumann called him) became increasingly isolated, worked to formalize theology using logic, became convinced that discoveries of Leibniz from the 1600s had been suppressed, and in 1978, with his wife's health failing, died of starvation, suspicious of doctors and afraid of being poisoned.

He left us the legacy of undecidability, which we now realize affects not just esoteric issues about mathematics, but also all sorts of questions in science, engineering, medicine and more.

One might think of undecidability as a limitation to progress, but in many ways it is instead a sign of richness. For with it comes computational irreducibility, and the possibility for systems to build up behavior beyond what can be summarized by simple formulas. Indeed, my own work suggests that much of the complexity we see in nature has precisely this origin. And perhaps it is also the essence of how from deterministic underlying laws we can build up apparent free will.

In science and technology we have normally crafted our theories and devices by careful design. But thinking in the abstract computational terms pioneered by Gödel's methods we can imagine an alternative. For if we represent everything uniformly in terms of rules or programs, we can in principle just explicitly enumerate all possibilities.

In the past, though, nothing like this ever seemed even faintly sensible. For it was implicitly assumed that to create a program with interesting behavior would require explicit human design—or at least the efforts of something like natural selection. But when I started actually doing experiments and systematically running the simplest programs, what I found instead is that the computational universe is teeming with diverse and complex behavior.

Already there is evidence that many of the remarkable forms we see in biology just come from sampling this universe. And perhaps by searching the computational universe we may find—even soon—the ultimate underlying laws for our own physical universe. (To discover all their consequences, though, will still require irreducible computational work.)

Exploring the computational universe puts mathematics too into a new context. For we can also now see a vast collection of alternatives to the mathematics that we have ultimately inherited from the arithmetic and geometry of ancient Babylon. And for example, the axioms of basic logic, far from being something special, now just appear as roughly the 50,000th possibility. And mathematics, long a purely theoretical science, must adopt experimental methods.

The exploration of the computational universe seems destined to become a core intellectual framework in the future of science. And in technology the computational universe provides a vast new resource that can be searched and mined for systems that serve our increasingly complex purposes. It is undecidability that guarantees an endless frontier of surprising and useful material to find.

And so it is that from Gödel's abstruse theorem about mathematics has emerged what I believe will be the defining theme of science and technology in the twenty-first century.

Alan Turing

June 23, 2012

I never met Alan Turing; he died five years before I was born. But somehow I feel I know him well—not least because many of my own intellectual interests have had an almost eerie parallel with his.

And by a strange coincidence, Mathematica's "birthday" (June 23, 1988) is aligned with Turing's—so that today is not only the centenary of Turing's birth, but is also Mathematica's 24th birthday.

I think I first heard about Alan Turing when I was about eleven years old, right around the time I saw my first computer. Through a friend of my parents, I had gotten to know a rather eccentric old classics professor, who, knowing my interest in science, mentioned to me this "bright young chap named Turing" whom he had known during the Second World War.

One of the classics professor's eccentricities was that whenever the word "ultra" came up in a Latin text, he would repeat it over and over again, and make comments about remembering it. At the time, I didn't think much of it—though I did remember it. Only years later did I realize that "Ultra" was the codename for the British cryptanalysis effort at Bletchley Park during the war. In a very British way, the classics professor wanted to tell me something about it, without breaking any secrets. And presumably it was at Bletchley Park that he had met Alan Turing.

A few years later, I heard scattered mentions of Alan Turing in various British academic circles. I heard that he had done mysterious but important work in breaking German codes during the war. And I heard it claimed that after the war, he had been killed by British Intelligence. At the time, at least some of the British wartime cryptography effort was still secret, including Turing's role in it. I wondered why. So I asked around, and started hearing that perhaps Turing had invented codes that were still being used. (In reality, the continued secrecy seems to have been intended to prevent it being known that certain codes had been broken—so other countries would continue to use them.)

I'm not sure where I next encountered Alan Turing. Probably it was when I decided to learn all I could about computer science—and saw all sorts of mentions of "Turing machines". But I have a distinct memory from around 1979 of going to the library and finding a little book about Alan Turing written by his mother, Sara Turing.

And gradually I built up quite a picture of Alan Turing and his work. And over the 30+ years that have followed, I have kept on running into Alan Turing, often in unexpected places.

In the early 1980s, for example, I had become very interested in theories of biological growth—only to find (from Sara Turing's book) that Alan Turing had done all sorts of largely unpublished work on that.

And for example in 1989, when we were promoting an early version of Mathematica, I decided to make a poster of the Riemann zeta function—only to discover that Alan Turing had at one time held the record for computing zeros of the zeta function. (Earlier he had also designed a gear-based machine for doing this.)

Recently I even found out that Turing had written about the "reform of mathematical notation and phraseology"—a topic of great interest to me in connection with both Mathematica and Wolfram|Alpha.

And at some point I learned that a high school math teacher of mine (Norman Routledge) had been a friend of Turing's late in his life. But even though my teacher knew my interest in computers, he never mentioned Turing or his work to me. And indeed, 35 years ago, Alan Turing and his work were little known, and it is only fairly recently that Turing has become as famous as he is today.

Turing's greatest achievement was undoubtedly his construction in 1936 of a universal Turing machine—a theoretical device intended to represent the mechanization of mathematical processes. And in some sense, Mathematica is precisely a concrete embodiment of the kind of mechanization that Turing was trying to represent.

In 1936, however, Turing's immediate purpose was purely theoretical. Indeed, it was to show not what could be mechanized in mathematics, but what could not. In 1931, Gödel's theorem had shown that there were limits to what could be proved in mathematics, and Turing wanted to understand the boundaries of what could ever be done by any systematic procedure in mathematics.

Turing was a young mathematician in Cambridge, England, and his work was couched in terms of mathematical problems of his time. But one of his steps was the theoretical construction of a universal Turing machine capable of being "programmed" to emulate any other Turing machine. In effect, Turing had invented the idea of universal computation—which was later to become the foundation on which all of modern computer technology is built.

At the time, though, Turing's work did not make much of a splash, probably largely because the emphasis of Cambridge mathematics was elsewhere. Just before Turing published his paper, he learned about a similar result by Alonzo Church from Princeton, formulated not in terms of theoretical machines, but in terms of the mathematics-like lambda calculus. And as a result Turing went to Princeton for a year to study with Church—and while he was there, wrote the most abstruse paper of his life.

The next few years for Turing were dominated by his wartime cryptanalysis work. I learned a few years ago that during the war Turing visited Claude Shannon at Bell Labs in connection with speech encipherment. Turing had been working on a kind of statistical approach to cryptanalysis—and I am extremely curious to know whether Turing told Shannon about this, and potentially launched the idea of information theory, which itself was first formulated for secret cryptanalysis purposes.

After the war, Turing got involved with the construction of the first actual computers in England. To a large extent, these computers had emerged from engineering, not from a fundamental understanding of Turing's work on universal computation.

There was, however, a definite, if circuitous, connection. In 1943 Warren McCulloch and Walter Pitts in Chicago wrote a theoretical paper about neural networks that used the idea of universal Turing machines to discuss general computation in the brain. John von Neumann read this paper, and used it in his recommendations about how practical computers should be built and programmed. (John von Neumann had known about Turing's paper in 1936, but at the time did not recognize its significance, instead describing Turing in a recommendation letter as having done interesting work on the central limit theorem.)

It is remarkable that in just over a decade Alan Turing was transported from writing theoretically about universal computation, to being able to write programs for an actual computer. I have to say, though, that from today's vantage point, his programs look incredibly "hacky"—with lots of special features packed in, and encoded as strange strings of letters. But perhaps to reach the edge of a new technology it's inevitable that there has to be hackiness.

And perhaps too it required a certain hackiness to construct the very first universal Turing machine. The concept was correct, but Turing quickly published an erratum to fix some bugs, and in later years, it's become clear that there were more bugs. But at the time Turing had no intuition about how easily bugs can occur.

Turing also did not know just how general or not his results about universal computation might be. Perhaps the Turing machine was just one model of a computational process, and other models—or brains—might have quite different capabilities. But gradually over the course of several decades, it became clear that a wide range of possible models were actually exactly equivalent to the machines Turing had invented.

It's strange to realize that Alan Turing never appears to have actually simulated a Turing machine on a computer. He viewed Turing machines as theoretical devices relevant for proving general principles. But he does not appear to have thought about them as concrete objects to be explicitly studied.

And indeed, when Turing came to make models of biological growth processes, he immediately started using differential equations—and appears never to have considered the possibility that something like a Turing machine might be relevant to natural processes.

When I became interested in simple computational processes around 1980, I also didn't consider Turing machines—and instead started off studying what I later learned were called cellular automata. And what I discovered was that even cellular automata with incredibly simple rules could produce incredibly complex behavior—which I soon realized could be considered as corresponding to a complex computation.

I probably simulated my first explicit Turing machine only in 1991. To me, Turing machines were built a little bit too much like engineering systems—and not like something that would likely correspond to a system in nature. But I soon found that even simple Turing machines, just like simple cellular automata, could produce immensely complex behavior.

In a sense, Alan Turing could easily have discovered this. But his intuition—like my original intuition—would have told him that no such phenomenon was possible. So it would likely only have been luck—and access to easy computation—that would have led him to find the phenomenon.

Had he done so, I am quite sure he would have become curious about just what the threshold for his concept of universality would be, and just how simple a Turing machine would suffice. In the mid-1990s, I searched the space of simple Turing machines, and found the smallest possible candidate. And after I put up a $25,000 prize, in 2007 Alex Smith showed that indeed this Turing machine is universal.

No doubt Alan Turing would quite quickly have grasped the significance of such results for thinking about both natural processes and mathematics. But without the empirical discoveries, his thinking did not progress in this direction.

Instead, he began to consider from a more engineering point of view to what extent computers should be able to emulate brains, and he invented ideas like the Turing test. Reading through his writings today, it is remarkable how many of his conceptual arguments about artificial intelligence still need to be made—though some, like his discussion of extrasensory perception, have become quaintly dated.

And looking at his famous 1950 article on "Computing Machinery and Intelligence" one sees a discussion of programming into a machine the contents of *Encyclopædia Britannica*—which he estimates should take 60 workers 50 years. I wonder what Alan Turing would think of Wolfram|Alpha, which, thanks to progress over the past 60 years, and perhaps some cleverness, has so far taken at least slightly less human effort.

In addition to his intellectual work, Turing has in recent times become something of a folk hero, most notably through the story of his death. Almost certainly it will never be known for sure whether his death was in fact intentional. But from what I know and have heard I must say that I rather doubt that it was.

When one first hears that Alan Turing died by eating an apple impregnated with cyanide one assumes it must have been intentional suicide. But when one later discovers that he was quite a tinkerer, had recently made cyanide for the purpose of electroplating spoons, kept chemicals alongside his food, and was rather a messy individual, the picture becomes a lot less clear.

I often wonder what Alan Turing would have been like to meet. I do not know of any recording of his voice (though he did once do a BBC radio broadcast). But I gather that even near the end of his life he giggled a lot, and talked with a kind of stutter that seemed to come from thinking faster than he was talking. He seemed to have found it easiest to talk to mathematicians. He thought a little about physics, though doesn't seem to have ever gotten deeply into it. And he seemed to have maintained a child-like enthusiasm and wonder for many intellectual questions throughout his life.

He was something of a loner, working successively on his own on his various projects. He was gay, and lived alone. He was no organizational politician, and towards the end of his life seems to have found himself largely ignored both by people working on computers and by people working on his new interest of biological growth and morphogenesis.

He was in some respects a quintessential British amateur, dipping his intellect into different areas. He achieved a high level of competence in pure mathematics, and used that as his professional base. His contributions in traditional mathematics were certainly perfectly respectable, though not spectacular. But in every area he touched, there was a certain crispness to the ideas he developed—even if their technical implementation was sometimes shrouded in arcane notation and masses of detail.

In some ways he was fortunate to live when he did. For he was at the right time to be able take the formalism of mathematics as it had been developed, and to combine it with the emerging engineering of his day, to see for the first time the general concept of computation.

It is perhaps a shame that he died 25 years before computer experiments became widely feasible. I certainly wonder what he would have discovered tinkering with Mathematica. I don't doubt that he would have pushed it to its limits, writing code that would horrify me. But I fully expect that long before I did, he would have discovered the main elements of *A New Kind of Science*, and begun to understand their significance.

He would probably be disappointed that 60 years after he invented the Turing test, there is still no full human-like artificial intelligence. And perhaps long ago he would have begun to campaign for the creation of something like Wolfram|Alpha, to turn human knowledge into something computers can handle.

If he had lived a few decades longer, he would no doubt have applied himself to a half dozen more areas. But there is still much to be grateful for in what Alan Turing did achieve in his 41 years, and his modern reputation as the founding father of the concept of computation—and the conceptual basis for much of what I, for example, have done—is well deserved.

John von Neumann

December 28, 2003

It would have been John von Neumann's 100th birthday today—if he had not died at age 54 in 1957. I've been interested in von Neumann for many years—not least because his work touched on some of my most favorite topics. He is mentioned in 12 separate places in *A New Kind of Science*—second in number only to Alan Turing, who appears 19 times.

I always feel that one can appreciate people's work better if one understands the people themselves better. And from talking to many people who knew him, I think I've gradually built up a decent picture of John von Neumann as a man.

He would have been fun to meet. He knew a lot, was very quick, always impressed people, and was lively, social and funny.

One video clip of him has survived. In 1955 he was on a television show called *Youth Wants to Know*, which today seems painfully hokey. Surrounded by teenage kids, he is introduced as a commissioner of the Atomic Energy Commission—which in those days was a big deal. He is asked about an exhibit of equipment. He says very seriously that it's mostly radiation detectors. But then a twinkle comes into his eye, and he points to another item, and says deadpan, "Except this, which is a carrying case." And that's the end of the only video record of John von Neumann that exists.

Some scientists (such as myself) spend most of their lives pursuing their own grand programs, ultimately in a fairly isolated way. John von Neumann was instead someone who always liked to interact with the latest popular issues—and the people around them—and then contribute to them in his own characteristic way.

He worked hard, often on many projects at once, and always seemed to have fun. In retrospect, he chose most of his topics remarkably well. He studied each of them with a definite practical mathematical style.

And partly by being the first person to try applying serious mathematical methods in various areas, he was able to make important and unique contributions.

But I've been told that he was never completely happy with his achievements because he thought he missed some great discoveries. And indeed he was close to a remarkable number of important mathematics-related discoveries of the twentieth century: Gödel's theorem, Bell's inequalities, information theory, Turing machines, computer languages—as well as my own more recent favorite core *A New Kind of Science* discovery of complexity from simple rules.

But somehow he never quite made the conceptual shifts that were needed for any of these discoveries.

There were, I think, two basic reasons for this. First, he was so good at getting new results by the mathematical methods he knew that he was always going off to get more results, and never had a reason to pause and see whether some different conceptual framework should be considered. And second, he was not particularly one to buck the system: he liked the social milieu of science and always seemed to take both intellectual and other authority seriously.

By all reports, von Neumann was something of a prodigy, publishing his first paper (on zeros of polynomials) at the age of 19. By his early twenties, he was established as a promising young professional mathematician—working mainly in the then-popular fields of set theory and foundations of math. (One of his achievements was alternate axioms for set theory.)

Like many good mathematicians in Germany at the time, he worked on David Hilbert's program for formalizing mathematics, and for example wrote papers aimed at finding a proof of consistency for the axioms of arithmetic. But he did not guess the deeper point that Kurt Gödel discovered in 1931: that actually such a proof is fundamentally impossible. I've been told that von Neumann was always disappointed that he had missed Gödel's theorem. He certainly knew all the methods needed to establish it (and understood it remarkably quickly once he heard it from Gödel). But somehow he did not have the brashness to disbelieve Hilbert, and go looking for a counterexample to Hilbert's ideas.

In the mid-1920s formalization was all the rage in mathematics, and quantum mechanics was all the rage in physics. And in 1927 von Neumann set out to bring these together—by axiomatizing quantum mechanics. A fair bit of the formalism that von Neumann built has become the standard framework for any mathematically oriented exposition of quantum mechanics. But I must say that I have always thought that it gave too much of an air of mathematical definiteness to ideas (particularly about quantum measurement) that in reality depend on all sorts of physical details. And indeed some of von Neumann's specific axioms turned out to be too restrictive for ordinary quantum mechanics—obscuring for a number of years the phenomenon of entanglement, and later of criteria such as Bell's inequalities.

But von Neumann's work on quantum mechanics had a variety of fertile mathematical spinoffs, and particularly what are now called von Neumann algebras have recently become popular in mathematics and mathematical physics.

Interestingly, von Neumann's approach to quantum mechanics was at first very much aligned with traditional calculus-based mathematics—investigating properties of Hilbert spaces, continuous operators, etc. But gradually it became more focused on discrete concepts, particularly early versions of "quantum logic". In a sense von Neumann's quantum logic ideas were an early attempt at defining a computational model of physics. But he did not pursue this, and did not go in the directions that have for example led to modern ideas of quantum computing.

By the 1930s von Neumann was publishing several papers a year, on a variety of popular topics in mainstream mathematics, often in collaboration with contemporaries of significant later reputation (Wigner, Koopman, Jordan, Veblen, Birkhoff, Kuratowski, Halmos, Chandrasekhar, etc.). Von Neumann's work was unquestionably good and innovative, though very much in the flow of development of the mathematics of its time.

Despite von Neumann's early interest in logic and the foundations of math, he (like most of the math community) moved away from this by the mid-1930s. In Cambridge and then in Princeton he encountered the young Alan Turing—even offering him a job as an assistant in 1938. But he apparently paid little attention to Turing's classic 1936 paper on Turing machines and the concept of universal computation, writing in a recommendation letter on June 1, 1937 that "[Turing] has done good work on ... theory of almost periodic functions and theory of continuous groups".

As it did for many scientists, von Neumann's work on the Manhattan Project appears to have broadened his horizons, and seems to have spurred his efforts to apply his mathematical prowess to problems of all sorts—not just in traditional mathematics. His pure mathematical colleagues seem to have viewed such activities as a peculiar and somewhat suspect hobby, but one that could generally be tolerated in view of his respectable mathematical credentials.

At the Institute for Advanced Study in Princeton, where von Neumann worked, there was strain, however, when he started a project to build an actual computer there. Indeed, even when I worked at the Institute in the early 1980s, there were still pained memories of the project. The pure mathematicians at the Institute had never been keen on it, and the story was that when von Neumann died, they had been happy to accept Thomas Watson of IBM's offer to send a truck to take away all of von Neumann's equipment. (Amusingly, the 6-inch on-off switch for the computer was left behind, bolted to the wall of the building, and has recently become a prized possession of a computer industry acquaintance of mine.)

I had some small interaction with von Neumann's heritage at the Institute in 1982 when the then-director (Harry Woolf) was recruiting me. (Harry's original concept was to get me to start a School of Computation at the Institute, to go along with the existing School of Natural Sciences and School of Mathematics. But for various reasons, this was not what happened.) I was concerned about intellectual property issues, having just had some difficulty with them at Caltech. Harry's response—that he attributed to the chairman of their board of trustees—was, "Look, von Neumann developed the computer here, but we insisted on giving it away; after that, why should we worry about any intellectual property rights?" (The practical result was a letter disclaiming any rights to any intellectual property that I produced at the Institute.)

Among several of von Neumann's interests outside of mainstream pure mathematics was his attempt to develop a mathematical theory of biology and life. In the mid-1940s there had begun to be—particularly from wartime work on electronic control systems—quite a bit of discussion about analogies between "natural and artificial automata", and "cybernetics". And von Neumann decided to apply his mathematical methods to this. I've been told he was particularly impressed by the work of McCullough and Pitts on formal models of the analogy between brains and electronics. (There were undoubtedly other influences too: John McCarthy told me that around 1948 he visited von Neumann, and told him about applying information theory ideas to thinking about the brain as an automaton; von Neumann's main response at the time was just, "Write it up!")

Von Neumann was in many ways a traditional mathematician, who (like Turing) believed he needed to turn to partial differential equations in describing natural systems. I've been told that at Los Alamos von Neumann was very taken with electrically stimulated jellyfish, which he appears to have viewed as doing some kind of continuous analog of the information processing of an electronic circuit. In any case, by about 1947, he had conceived the idea of using partial differential equations to model a kind of factory that could reproduce itself, like a living organism.

Von Neumann always seems to have been very taken with children, and I am told that it was in playing with an erector set owned by the son of his game-theory collaborator Oskar Morgenstern that von Neumann realized that his self-reproducing factory could actually be built out of discrete robotic-like parts. (There was already something of a tradition of building computers out of Meccano—and indeed for example some of Hartree's early articles on analog computers appeared in *Meccano Magazine*.)

An electrical engineer named Julian Bigelow, who worked on von Neumann's IAS computer project, pointed out that 3D parts were not necessary, and that 2D would work just as well. (When I was at the Institute in the early 1980s Bigelow was still there, though unfortunately viewed as a slightly peculiar relic of von Neumann's project.)

Stan Ulam told me that he had independently thought about making mathematical models of biology, but in any case, around 1951 he appears to have suggested to von Neumann that one should be able to use a simplified, essentially combinatorial model—based on something like the infinite matrices that Ulam had encountered in the so-called Scottish Book of math problems (named after a café in Poland) to which he had contributed.

The result of all this was a model that was formally a two-dimensional cellular automaton. Systems equivalent to two-dimensional cellular automata were arising in several other contexts around the same time (see *A New Kind of Science*). Von Neumann seems to have viewed his version as a convenient framework in which to construct a mathematical system that could emulate engineered computer systems—especially the EDVAC on which von Neumann worked.

In the period 1952–53 von Neumann sketched an outline of a proof that it was possible for a formal system to support self reproduction. Whenever he needed a different kind of component (wire, oscillator, logic element, etc.) he just added it as a new state of his cellular automaton, with new rules. He ended up with a 29-state system, and a 200,000-cell configuration that could reproduce itself. (Von Neumann himself did not complete the construction. This was done in the early 1960s by a former assistant of von Neumann's named Arthur Burks, who had left the IAS computer project to concentrate on his interests in philosophy, though who maintains even today an interest in cellular automata.)

From the point of view of *A New Kind of Science*, von Neumann's system now seems almost grotesquely complicated. But von Neumann's intuition told him that one could not expect a simpler system to show something as sophisticated and biological as self reproduction. What he said was that he thought that below a certain level of complexity, systems

---

Arthur Burks died in 2008, at the age of 92.

would always be "degenerative", and always generate what amounts to behavior simpler than their rules. But then, from seeing the example of biology, and of systems like Turing machines, he believed that above some level, there should be an "explosive" increase in complexity, with systems able to generate other systems more complex than themselves. But he said that he thought the threshold for this would be systems with millions of parts.

Twenty-five years ago I might not have disagreed too strongly with that. And certainly for me it took several years of computer experimentation to understand that in fact it takes only very simple rules to produce even the most complex behavior. So I do not think it surprising—or unimpressive—that von Neumann failed to realize that simple rules were enough.

Of course, as one often realizes in retrospect, he did have some other clues. He knew about the idea of generating pseudorandom numbers from simple rules, suggesting the so-called "middle square method". He had the beginnings of the idea of doing computer experiments in areas like number theory. He analyzed the first 2000 digits of π and e, computed on the ENIAC, finding that they seemed random—though making no comment on it. (He also looked at ContinuedFraction[$2^{1/3}$].)

I have asked many people who knew him why von Neumann never considered simpler rules. Marvin Minsky told me that he actually asked von Neumann about this directly, but that von Neumann had been somewhat confused by the question. It would have been much more Ulam's style than von Neumann's to have come up with simpler rules, and Ulam indeed did try making a one-dimensional analog of 2D cellular automata, but came up not with 1D cellular automata, but with a curious number-theoretical system.

In the last ten years of his life, von Neumann got involved in an impressive array of issues. Some of his colleagues seem to have felt that he spent too little time on each one, but still his contributions were usually substantial—sometimes directly in terms of content, and usually at least in terms of lending his credibility to emerging areas.

He made mistakes, of course. He thought that each logical step in computation would necessarily dissipate a certain amount of heat, whereas in fact reversible computation is in principle possible. He thought that the unreliability of components would be a major issue in building large computer systems; he apparently did not have an idea like error-correcting codes. He is reputed to have said that no computer program would ever be more than a few thousand lines long. He was probably thinking about proofs of theorems—but did not think about subroutines, the analog of lemmas.

Von Neumann was a great believer in the efficacy of mathematical methods and models, perhaps implemented by computers. In 1950 he was optimistic that accurate numerical weather forecasting would soon be possible. In addition, he believed that with methods like game-theory it should be possible to understand much of economics and other forms of human behavior.

Von Neumann was always quite a believer in using the latest methods and tools (I'm sure he would have been a big Mathematica user today). He typically worked directly with one or two collaborators, sometimes peers, sometimes assistants, though he maintained contact with a large network of scientists. (A typical communication was a letter he wrote to Alan Turing in 1949, in which he asks, "What are the problems on which you are working now, and what is your program for the immediate future?") In his later years he often operated as a distinguished consultant, brought in by the government, or other large organizations. His work was then often presented as a report, that was accorded particular weight because of his distinguished consultant status. (It was also often a good and clear piece of work.) He was often viewed a little ambivalently as an outsider in the fields he entered—positively because he brought his distinction to the field, negatively because he was not in the clique of experts in the field.

Particularly in the early 1950s, von Neumann became deeply involved in military consulting, and indeed I wonder how much of the intellectual style of Cold War US military strategic thinking actually originated with him. He seems to have been quite flattered that he was called upon to do this consulting, and he certainly treated the government with considerably more respect than many other scientists of his day. Except sometimes in his exuberance to demonstrate his mathematical and calculational prowess, he seems to have always been quite mature and diplomatic. The transcript of his testimony at the Oppenheimer security hearing certainly for example bears this out.

Nevertheless, von Neumann's military consulting involvements left some factions quite negative about him. It's sometimes said, for example, that von Neumann might have been the model for the sinister Dr. Strangelove character in Stanley Kubrick's movie of that name (and indeed von Neumann was in a wheelchair for the last year of his life). And vague negative feelings about von Neumann surface for example in a typical statement I heard recently from a science historian of the period—that "somehow I don't like von Neumann, though I can't remember exactly why".

I recently met von Neumann's only child—his daughter Marina, who herself has had a distinguished career, mostly at General Motors. She reinforced my impression that until his unpleasant final illness, John von Neumann was a happy and energetic man, working long hours on mathematical topics, and always having fun. She told me that when he died, he left a box that he directed should be opened fifty years after his death. What does it contain? His last sober predictions of a future we have now seen? Or a joke—like a funny party hat of the type he liked to wear? It'll be most interesting in 2007 to find out.

John von Neumann's Box

February 8, 2007

At the end of my piece about the 100th anniversary of John von Neumann's birth, I mentioned that his daughter had told me about a box to be opened on the 50th anniversary of his death. That anniversary is today.

And being reminded of this last week, I sent mail to his daughter to ask what had become of the box.

Disappointingly, she responded, "The Big Box opening turned out to be a Big Bust." Apparently she and her children and their grandchildren had all assembled ... only to discover that it was all a big mistake: the box was actually not John von Neumann's at all!

Perhaps it was for the best. Von Neumann's daughter sent me a piece she wrote about him, pointing out the unpredictability (irreducibility?) of technological and other change—and expressing concern that our species might wipe itself out by the year 1980.

Fortunately that of course hasn't happened. And it was recently suggested to me that perhaps The Box might contain some Cold-War-inspired plan for a Dr.-Strangelove-like Doomsday Machine. So—"von Neumann machine" self-replicators notwithstanding—it's perhaps just as well that in the end there was no box.

George Boole

November 2, 2015

Today is the 200th anniversary of the birth of George Boole. In our modern digital world, we're always hearing about "Boolean variables"—1 or 0, true or false. And one might think, "What a trivial idea! Why did someone even explicitly need to invent it?" But as is so often the case, there's a deeper story—for Boolean variables were really just a side effect of an important intellectual advance that George Boole made.

When George Boole came onto the scene, the disciplines of logic and mathematics had developed quite separately for more than 2000 years. And George Boole's great achievement was to show how to bring them together, through the concept of what's now called Boolean algebra. And in doing so he effectively created the field of mathematical logic, and set the stage for the long series of developments that led for example to universal computation.

When George Boole invented Boolean algebra, his basic goal was to find a set of mathematical axioms that could reproduce the classical results of logic. His starting point was ordinary algebra, with variables like x and y, and operations like addition and multiplication.

At first, ordinary algebra seems a lot like logic. After all, *p and q* is the same as *q and p*, just as $p \times q = q \times p$. But if one looks in more detail, there are differences. Like $p \times p = p^2$, but *p and p* is just p. Somewhat confusingly, Boole used the notation of standard algebra, but added special rules to create an axiom system that he then showed could reproduce all the usual results of logic.

Boole was rather informal in the way he described his axiom system. But within a few decades, it had been more precisely formalized, and over the course of the century that followed, a few progressively simpler forms of it were found. And then, as it happens, 16 years ago I ended up finishing this 150-year process, by finding—largely as a side effect of other science I was doing—the provably very simplest possible axiom system for logic, that actually happens to consist of just a single axiom.

Year	Axioms
1847 1904	$p \times q = q \times p$
	$p + q = q + p$
	$p \times (q + (-q)) = p$
	$p + (q \times -q) = p$
	$p \times (q + r) = (p \times q) + (p \times r)$
	$p + (q \times r) = (p + q) \times (p + r)$
1999	$((p \cdot q) \cdot r) \cdot (p \cdot ((p \cdot r) \cdot p)) = r$

Year	Axioms
1913	$(p \cdot p) \cdot (p \cdot p) = p$
	$p \cdot (q \cdot (q \cdot q)) = p \cdot p$
	$(p \cdot (q \cdot r)) \cdot (p \cdot (q \cdot r)) = ((q \cdot q) \cdot p) \cdot ((r \cdot r) \cdot p)$
1933	$p + q = q + p$
	$p + (q + r) = (p + q) + r$
	$-(-p + q) + -((-p) + (-q)) = p$
1949	$(p \cdot (q \cdot r)) \cdot (p \cdot (q \cdot r)) = ((r \cdot p) \cdot p) \cdot ((q \cdot p) \cdot p)$
	$(p \cdot p) \cdot (q \cdot p) = p$

× And + Or − Not • Nand

I thought this axiom was pretty neat, and looking at where it lies in the space of possible axioms has interesting implications for the foundations of mathematics and logic. But in the context of George Boole, one can say that it's a minimal version of his big idea: that one can have a mathematical axiom system that reproduces all the results of logic just by what amount to simple algebra-like transformations.

Who Was George Boole?

But let's talk about George Boole, the person. Who was he, and how did he come to do what he did?

George Boole was born in 1815, in England, in the fairly small town of Lincoln, about 120 miles north of London. His father had a serious interest in science and mathematics, and had a small business as a shoemaker. George Boole was something of a self-taught prodigy, who first became locally famous at age 14 with a translation of a Greek poem that he published in the local newspaper. At age 16 he was hired as a teacher at a local school, and by that time he was reading calculus books, and apparently starting to formulate what would later be his idea about relations between mathematics and logic.

At age 19, George Boole did a startup: he started his own elementary school. It seems to have been decently successful, and in fact Boole continued making his living running (or "conducting" as it was then called) schools until he was in his thirties. He was involved with a few people educated in places like Cambridge, notably through the local Mechanics' Institute (a little like a modern community college). But mostly he seems just to have learned by reading books on his own.

He took his profession as a schoolteacher seriously, and developed all sorts of surprisingly modern theories about the importance of understanding and discovery (as opposed to rote memorization), and the value of tangible examples in areas like mathematics (he surely would have been thrilled by what's now possible with computers).

When he was 23, Boole started publishing papers on mathematics. His early papers were about hot topics of the time, such as calculus of variations. Perhaps it was his interest in education and exposition that led him to try creating different formalisms, but soon he became a pioneer in the "calculus of operations": doing calculus by manipulating operators rather than explicit algebraic expressions.

It wasn't long before he was interacting with leading British mathematicians of the day, and getting positive feedback. He considered going to Cambridge to become a "university person", but was put off when told that he would have to start with the standard undergraduate course, and stop doing his own research.

Mathematical Analysis of Logic

Logic as a field of study had originated in antiquity, particularly with the work of Aristotle. It had been a staple of education throughout the Middle Ages and beyond, fitting into the practice of rote learning by identifying specific patterns of logical arguments ("syllogisms") with mnemonics like "bArbArA" and "cElArEnt". In many ways, logic hadn't changed much in over a thousand years, though by the 1800s there were efforts to make it more streamlined and "formal". But the question was how. And in particular, should this happen through the methods of philosophy, or mathematics?

In early 1847, Boole's friend Augustus De Morgan had become embroiled in a piece of academic unpleasantness over the question. And this led

Boole quickly to go off and work out his earlier ideas about how logic could be formulated using mathematics. The result was his first book, *The Mathematical Analysis of Logic*, published the same year:

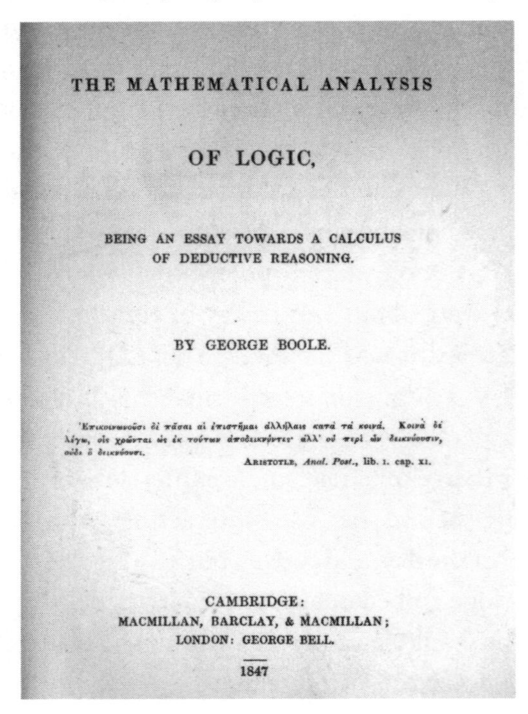

THE MATHEMATICAL ANALYSIS

OF LOGIC,

BEING AN ESSAY TOWARDS A CALCULUS
OF DEDUCTIVE REASONING.

BY GEORGE BOOLE.

Ἐπικοινωνοῦσι δέ πᾶσαι αἱ ἐπιστῆμαι ἀλλήλαις κατά τά κοινά. Κοινὰ δέ λέγω, οἷς χρῶνται ὡς ἐκ τούτων ἀποδεικνύντες· ἀλλ' οὐ περὶ ὧν δεικνύουσιν, οὐδὲ ὃ δεικνύουσι.

ARISTOTLE, *Anal. Post.*, lib. I. cap. XI.

CAMBRIDGE:
MACMILLAN, BARCLAY, & MACMILLAN;
LONDON: GEORGE BELL.

1847

The book was not long—only 86 pages. But it explained Boole's idea of representing logic using a form of algebra. The notion that one could have an algebra with variables that weren't just ordinary numbers happened to have just arisen in Hamilton's 1843 invention of quaternion algebra—and Boole was influenced by this. (Galois had also done something similar in 1832 working on groups and finite fields.)

150 years before Boole, Gottfried Leibniz had also thought about using algebra to represent logic. But he'd never managed to see quite how. And the idea seems to have been all but forgotten until Boole finally succeeded in doing it in 1847.

Looking at Boole's book today, much of it is quite easy to understand. Here, for example, is him showing how his algebraic formulation reproduces a few standard results in logic:

1st. Disjunctive Syllogism.

Either X is true, or Y is true (exclusive), $x + y - 2xy = 1$
But X is true, $x = 1$
Therefore Y is not true, $\therefore y = 0$

Either X is true, or Y is true (not exclusive), $x + y - xy = 1$
But X is not true, $x = 0$
Therefore Y is true, $\therefore y = 1$

2nd. Constructive Conditional Syllogism.

If X is true, Y is true, $x(1 - y) = 0$
But X is true, $x = 1$
Therefore Y is true, $\therefore 1 - y = 0$ or $y = 1$.

3rd. Destructive Conditional Syllogism.

If X is true, Y is true, $x(1 - y) = 0$
But Y is not true, $y = 0$
Therefore X is not true, $\therefore x = 0$

4th. Simple Constructive Dilemma, the minor premiss exclusive.

If X is true, Y is true, $x(1 - y) = 0$, (41),
If Z is true, Y is true, $z(1 - y) = 0$, (42),
But Either X is true, or Z is true, $x + z - 2xz = 1$, (43).

From the equations (41), (42), (43), we have to eliminate x and z. In whatever way we effect this, the result is

$$y = 1;$$

whence it appears that the Proposition Y is true.

At a surface level, this all seems fairly straightforward. "And" is represented by multiplication of variables xy, "not" by $1-x$, and "(exclusive) or" by $x+y-2xy$. There are also extra constraints like $x^2=x$. But when one tries digging deeper, things become considerably murkier. Just what are x and y supposed to be? Today we'd call these Boolean variables, and imagine they could have discrete values 1 or 0, representing true or false. But Boole seems to have never wanted to talk about anything that explicit, or anything discrete or combinatorial. All he ever seemed to discuss was algebraic expressions and equations—even to the point of using series expansions to effectively enumerate possible combinations of values for logical variables.

The Laws of Thought

When Boole wrote his first book he was still working as a teacher and running a school. But he had also become well known as a mathematician, and in 1849, when Queen's College, Cork (now University College Cork) opened in Ireland, Boole was hired as its first math professor. And once in Cork, Boole started to work on what would become his most famous book, *An Investigation of the Laws of Thought*:

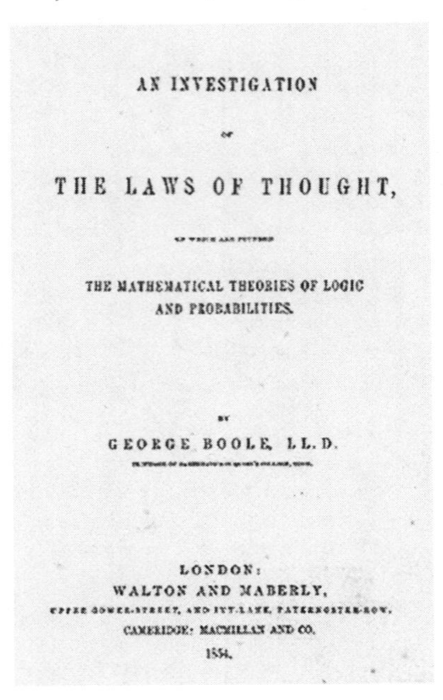

AN INVESTIGATION

OF

THE LAWS OF THOUGHT,

THE MATHEMATICAL THEORIES OF LOGIC
AND PROBABILITIES.

BY

GEORGE BOOLE, LL.D.

LONDON:
WALTON AND MABERLY,
UPPER GOWER-STREET, AND IVY-LANE, PATERNOSTER-ROW.
CAMBRIDGE: MACMILLAN AND CO.
1854.

His preface began, "The design of the following treatise is to investigate the fundamental laws of those operations of the mind by which reasoning is performed; to give expression to them in the symbolical language of a Calculus, and upon this foundation to establish the science of Logic and construct its method; ..."

Boole appears to have seen himself as trying to create a calculus for the "science of intellectual powers" analogous to Newton's calculus for physical science. But while Newton had been able to rely on concepts like space and time to inform the structure of his calculus, Boole had to build on the basis of a model of how the mind works, which for him was unquestionably logic.

The first part of *Laws of Thought* is basically a recapitulation of Boole's earlier book on logic, but with additional examples—such as a chapter covering logical proofs about the existence and characteristics of God. The second part of the book is in a sense more mathematically traditional. For instead of interpreting his algebraic variables as related to logic, he interprets them as traditional numbers corresponding to probabilities—and in doing so shows that the laws for combining probabilities of events have the same structure as the laws for combining logical statements.

For the most part *Laws of Thought* reads like a mathematical work, with abstract definitions and formal conclusions. But in the final chapter Boole tries to connect what he has done to empirical questions about the operation of the mind. He discusses how free will can be compatible with definite laws of thought. He talks about how imprecise human experiences can lead to precise concepts. He discusses whether there is truth that humans can recognize that goes beyond what mathematical laws can ever explain. And he talks about how an understanding of human thinking should inform education.

The Rest of Boole's Life

After the publication of *Laws of Thought*, George Boole stayed in Cork, living another decade and dying in 1864 of pneumonia at the age of 49. He continued to publish widely on mathematics, but never published on logic again, though he probably intended to do so.

In his lifetime, Boole was much more recognized for his work on traditional mathematics than on logic. He wrote two textbooks, one in 1859 on differential equations, and one in 1860 on difference equations. Both are clean and elegant expositions. And interestingly, while there are endless modern alternatives to Boole's *Differential Equations*, sufficiently little has been done on difference equations that when we were implementing them in Mathematica in the late 1990s, Boole's 1860 book was still an important reference, notable especially for its nice examples of the factorization of linear difference operators.

What Was Boole Like?

What was Boole like as a person? There's quite a bit of information on this, not least from his wife's writings and from correspondence and reminiscences his sister collected when he died. From what one can tell, Boole was organized and diligent, with careful attention to detail.

He worked hard, often late into the night, and could be so engrossed in his work that he became quite absent minded. Despite how he looks in pictures, he appears to have been rather genial in person. He was well liked as a teacher, and was a talented lecturer, though his blackboard writing was often illegible. He was a gracious and extensive correspondent, and made many visits to different people and places. He spent many years managing people, first at schools, and then at the university in Cork. He had a strong sense of justice, and while he did not like controversy, he was occasionally involved in it, and was not shy to maintain his position.

Despite his successes, Boole seems to have always thought of himself as a self-taught schoolteacher, rather than a member of the academic elite. And perhaps this helped in his ability to take intellectual risks. Whether it was playing fast and loose with differential operators in calculus, or finding ways to bend the laws of algebra so they could apply to logic, Boole seems to have always taken the attitude of just moving forward and seeing where he could go, trusting his own sense of what was correct and true.

Boole was single most of his life, though finally married at the age of 40. His wife, Mary Everest Boole, was 17 years his junior, and she outlived him by 52 years, dying in 1916. She had an interesting story in her own right, later in her life writing books with titles like *Philosophy and Fun of Algebra*, *Logic Taught by Love*, *The Preparation of the Child for Science* and *The Message of Psychic Science to the World*. George and Mary Boole had five daughters—who, along with their own children, had a wide range of careers and accomplishments, some quite mathematical.

Legacy

It is something of an irony that George Boole, committed as he was to the methods of algebra, calculus and continuous mathematics, should have come to symbolize discrete variables. But to be fair, this took a while. In the decades after he died, the primary influence of Boole's work on logic was on the wave of abstraction and formalization that swept through mathematics—involving people like Frege, Peano, Hilbert,

Whitehead, Russell and eventually Gödel and Turing. And it was only in 1937, with the work of Claude Shannon on switching networks, that Boolean algebra began to be used for practical purposes.

Today there is a lot on Boolean computation in Mathematica and the Wolfram Language, and in fact George Boole is the person with the largest number (15) of distinct functions in the system named after them.

But what has made Boole's name so widely known is not Boolean algebra, it's the much simpler notion of Boolean variables, which appear in essentially every computer language—leading to a progressive increase in mentions of the word "Boolean" in publications since the 1950s:

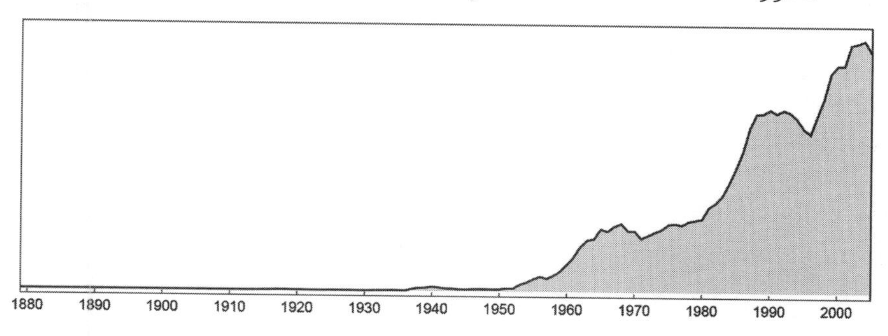

Was this inevitable? In some sense, I suspect it was. For when one looks at history, sufficiently simple formal ideas have a remarkable tendency to eventually be widely used, even if they emerge only slowly from quite complex origins. Most often what happens is that at some moment the ideas become relevant to technology, and quickly then go from curiosities to mainstream.

My work on *A New Kind of Science* has made me think about enumerations of what amount to all possible "simple formal ideas". Some have already become incorporated in technology, but many have not yet. But the story of George Boole and Boolean variables provides an interesting example of what can happen over the course of centuries—and how what at first seems obscure and abstruse can eventually become ubiquitous.

Ada Lovelace

December 10, 2015

Ada Lovelace was born 200 years ago today. To some she is a great hero in the history of computing; to others an overestimated minor figure. I've been curious for a long time what the real story is. And in preparation for her bicentennial, I decided to try to solve what for me has always been the "mystery of Ada".

It was much harder than I expected. Historians disagree. The personalities in the story are hard to read. The technology is difficult to understand. The whole story is entwined with the customs of 19th-century British high society. And there's a surprising amount of misinformation and misinterpretation out there.

But after quite a bit of research—including going to see many original documents—I feel like I've finally gotten to know Ada Lovelace, and gotten a grasp on her story. In some ways it's an ennobling and inspiring story; in some ways it's frustrating and tragic.

It's a complex story, and to understand it, we'll have to start by going over quite a lot of facts and narrative.

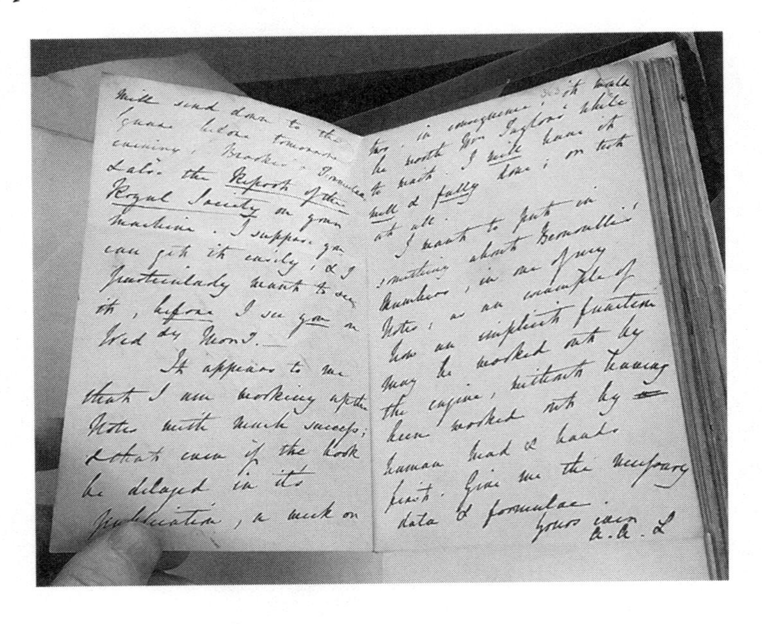
will send down to the Square before tomorrow evening; Brooker's Formulae & also the Report of the Royal Society on your machine. I suppose you can get it easily, & I particularly want to see it, before I see you on Wedy Morng.

It appears to me that I am working up the Notes with much success; & that even if the book be delayed in its publication, a week on

tory, in consequence, it would be worth your while to wait. I will have it well & fully done; or not at all.

I want to put in something about Bernoulli's Numbers, in one of my Notes, as an example of how an implicit function may be worked out by the engine, without having been worked out by human head & hands first. Give me the necessary data & formulae.

Yours ever
A. A. L

The Early Life of Ada

Let's begin at the beginning. Ada Byron, as she was then called, was born in London on December 10, 1815 to recently married high-society parents. Her father, Lord Byron (George Gordon Byron), was 27 years old, and had just achieved rock-star status in England for his poetry. Her mother, Annabella Milbanke, was a 23-year-old heiress committed to progressive causes, who inherited the title Baroness Wentworth. Her father said he gave her the name "Ada" because "It is short, ancient, vocalic".

Ada's parents were something of a study in opposites. Byron had a wild life—and became perhaps the top "bad boy" of the 19th century—with dark episodes in childhood, and lots of later romantic and other excesses. In addition to writing poetry and flouting the social norms of his time, he was often doing the unusual: keeping a tame bear in his college rooms in Cambridge, living it up with poets in Italy and "five peacocks on the grand staircase", writing a grammar book of Armenian, and—had he not died too soon—leading troops in the Greek war of independence (as celebrated by a big statue in Athens), despite having no military training whatsoever.

Annabella Milbanke was an educated, religious and rather proper woman, interested in reform and good works, and nicknamed by Byron "Princess of Parallelograms". Her very brief marriage to Byron fell apart when Ada was just 5 weeks old, and Ada never saw Byron again (though he kept a picture of her on his desk and famously mentioned her in his poetry). He died at the age of 36, at the height of his celebrityhood, when Ada was 8. There was enough scandal around him to fuel hundreds of books, and the PR battle between the supporters of Lady Byron (as Ada's mother styled herself) and of him lasted a century or more.

Ada led an isolated childhood on her mother's rented country estates, with governesses and tutors and her pet cat, Mrs. Puff. Her mother, often absent for various (quite wacky) health cures, enforced a system of education for Ada that involved long hours of study and exercises in self control. Ada learned history, literature, languages, geography, music, chemistry, sewing, shorthand and mathematics (taught in part through experiential methods) to the level of elementary geometry and algebra. When Ada was 11, she went with her mother and an entourage on a year-long tour of Europe. When she returned she was enthusiastically doing things like studying what she called "flyology"—and imagining how to mimic bird flight with steam-powered machines.

But then she got sick with measles (and perhaps encephalitis)—and ended up bedridden and in poor health for 3 years. She finally recovered in time to follow the custom for society girls of the period: on turning 17 she went to London for a season of socializing. On June 5, 1833, 26 days after she was "presented at Court" (i.e. met the king), she went to a party at the house of 41-year-old Charles Babbage (whose oldest son was the same age as Ada). Apparently she charmed the host, and he invited her and her mother to come back for a demonstration of his newly constructed Difference Engine: a 2-foot-high hand-cranked contraption with 2000 brass parts, now to be seen at the Science Museum in London:

Ada's mother called it a "thinking machine", and reported that it "raised several Nos. to the 2nd & 3rd powers, and extracted the root of a Quadratic Equation". It would change the course of Ada's life.

Charles Babbage

What was the story of Charles Babbage? His father was an enterprising and successful (if personally distant) goldsmith and banker. After various schools and tutors, Babbage went to Cambridge to study mathematics, but soon was intent on modernizing the way mathematics was done there, and with his lifelong friends John Herschel (son of the discoverer of Uranus) and George Peacock (later a pioneer in abstract algebra), founded the Analytical Society (which later became the Cambridge Philosophical Society) to push for reforms like replacing Newton's ("British") dot-based notation for calculus with Leibniz's ("Continental") function-based one.

Babbage graduated from Cambridge in 1814 (a year before Ada Lovelace was born), went to live in London with his new wife, and started establishing himself on the London scientific and social scene. He didn't have a job as such, but gave public lectures on astronomy and wrote respectable if unspectacular papers about various mathematical topics (functional equations, continued products, number theory, etc.)—and was supported, if modestly, by his father and his wife's family.

In 1819 Babbage visited France, and learned about the large-scale government project there to make logarithm and trigonometry tables. Mathematical tables were of considerable military and commercial significance in those days, being used across science, engineering and finance, as well as in areas like navigation. It was often claimed that errors in tables could make ships run aground or bridges collapse.

Back in England, Babbage and Herschel started a project to produce tables for their new Astronomical Society, and it was in the effort to check these tables that Babbage is said to have exclaimed, "I wish to God these tables had been made by steam!"—and began his lifelong effort to mechanize the production of tables.

State of the Art

There were mechanical calculators long before Babbage. Pascal made one in 1642, and we now know there was even one in antiquity. But in Babbage's day such machines were still just curiosities, not reliable enough for everyday practical use. Tables were made by human computers, with the work divided across a team, and the lowest-level computations being based on evaluating polynomials (say from series expansions) using the method of differences.

What Babbage imagined is that there could be a machine—a Difference Engine—that could be set up to compute any polynomial up to a certain degree using the method of differences, and then automatically step through values and print the results, taking humans and their propensity for errors entirely out of the loop.

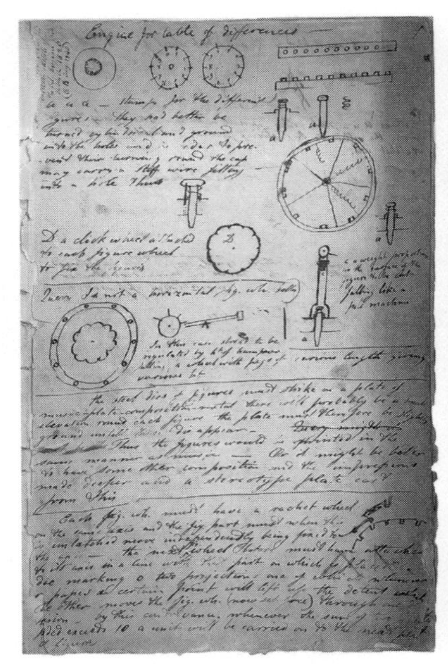

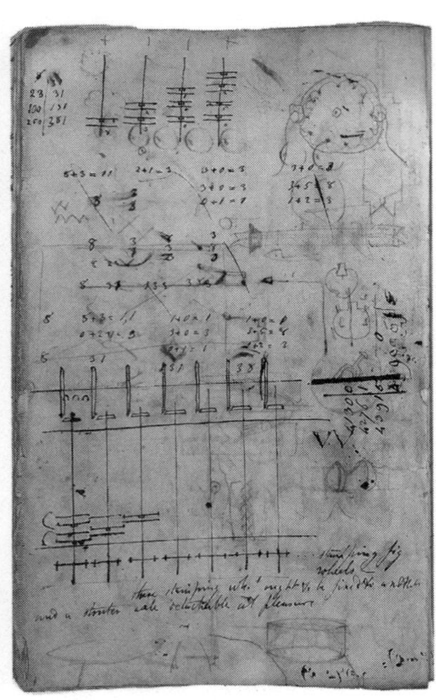

By early 1822, the 30-year-old Babbage was busy studying different types of machinery, and producing plans and prototypes of what the Difference Engine could be. The Astronomical Society he'd co-founded awarded him a medal for the idea, and in 1823 the British government agreed to provide funding for the construction of such an engine.

Babbage was slightly distracted in 1824 by the prospect of joining a life insurance startup, for which he did a collection of life-table calculations. But he set up a workshop in his stable (his "garage"), and kept on having ideas about the Difference Engine and how its components could be made with the tools of his time.

In 1827, Babbage's table of logarithms—computed by hand—was finally finished, and would be reprinted for nearly 100 years. Babbage had them printed on yellow paper on the theory that this would minimize user error. (When I was in elementary school, logarithm tables were still the fast way to do multiplication.)

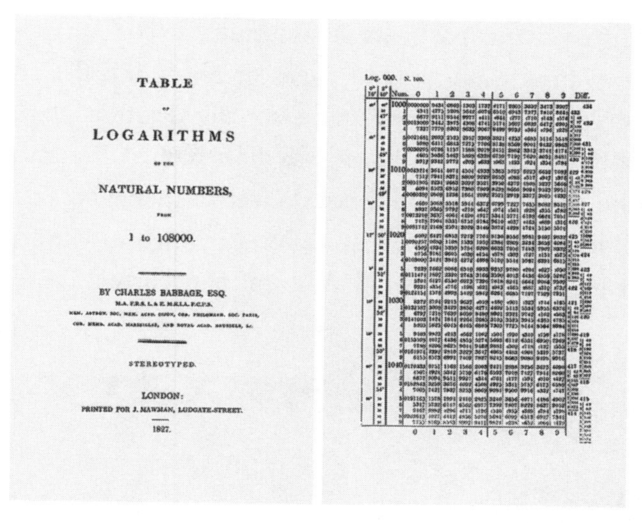

TABLE
OF
LOGARITHMS
OF THE
NATURAL NUMBERS,
FROM
1 to 108000.

BY CHARLES BABBAGE, ESQ.

STEREOTYPED.

LONDON:
PRINTED FOR J. MAWMAN, LUDGATE-STREET.
1827.

Also in 1827, Babbage's father died, leaving him about £100K, or perhaps $14 million today, setting up Babbage financially for the rest of his life. The same year, though, his wife died. She had had eight children with him, but only three survived to adulthood.

Dispirited by his wife's death, Babbage took a trip to continental Europe, and being impressed by what he saw of the science being done there, wrote a book entitled *Reflections on the Decline of Science in England*, that ended up being mainly a diatribe against the Royal Society (of which he was a member).

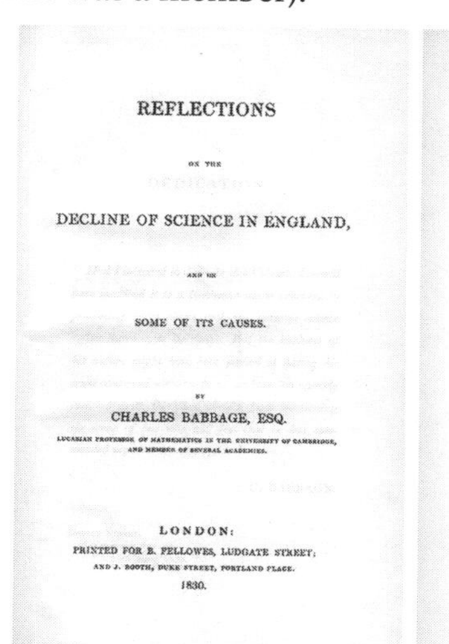

REFLECTIONS
ON THE
DECLINE OF SCIENCE IN ENGLAND,
AND ON
SOME OF ITS CAUSES.
BY
CHARLES BABBAGE, ESQ.
LUCASIAN PROFESSOR OF MATHEMATICS IN THE UNIVERSITY OF CAMBRIDGE, AND MEMBER OF SEVERAL ACADEMIES.

LONDON:
PRINTED FOR B. FELLOWES, LUDGATE STREET;
AND J. BOOTH, DUKE STREET, PORTLAND PLACE.
1830.

PREFACE.

Of the causes which have induced me to print this volume I have little to say; my own opinion is, that it will ultimately do some service to science, and without that belief I would not have undertaken so thankless a task. That it is too true not to make enemies, is an opinion in which I concur with several of my friends, although I should hope that what I have written will not give just reason for the permanence of such feelings. On one point I shall speak decidedly, it is not connected in any degree with the calculating machine on which I have been engaged; the causes which have led to it have been long operating, and would have produced this result whether I had ever speculated on that subject, and whatever might have been the fate of my speculations.

If any one shall endeavour to account for the opinions stated in these pages by ascribing them to any imagined circumstance peculiar to myself, I think he will be mistaken. That science has long been neglected and declining in England, is not an opinion originating with me, but is shared by many, and has been expressed by higher authority than mine. I shall offer a few notices on this subject, which, from their scattered position, are unlikely to have met the reader's

Though often distracted, Babbage continued to work on the Difference Engine, generating thousands of pages of notes and designs. He was quite hands on when it came to personally drafting plans or doing machine-shop experiments. But he was quite hands off in managing the engineers he hired—and he did not do well at managing costs. Still, by 1832 a working prototype of a small Difference Engine (without a printer) had successfully been completed. And this is what Ada Lovelace saw in June 1833.

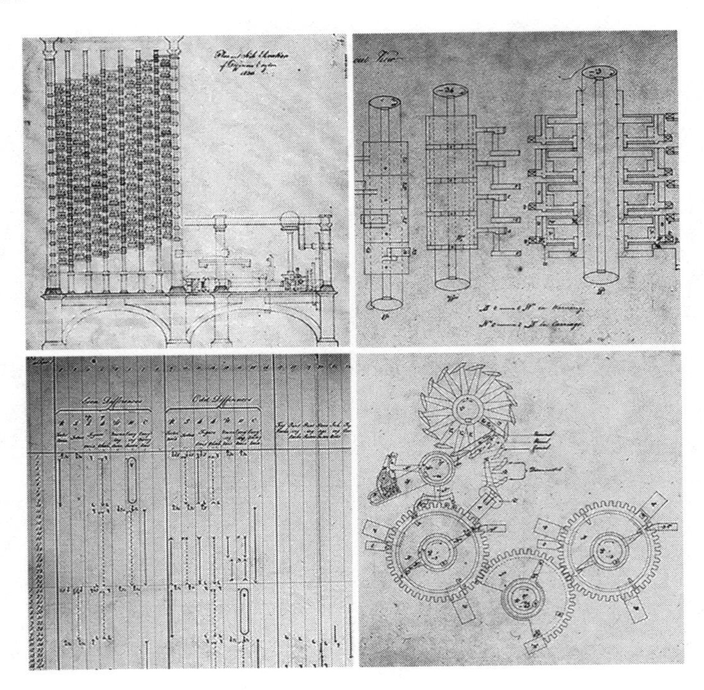

Back to Ada

Ada's encounter with the Difference Engine seems to be what ignited her interest in mathematics. She had gotten to know Mary Somerville, translator of Laplace and a well-known expositor of science—and partly with her encouragement, was soon, for example, enthusiastically studying Euclid. And in 1834, Ada went along on a philanthropic tour of mills in the north of England that her mother was doing, and was quite taken with the then-high-tech equipment they had.

On the way back, Ada taught some mathematics to the daughters of one of her mother's friends. She continued by mail, noting that this could be "the commencement of 'A Sentimental Mathematical Correspondence carried on for years between two ladies of rank' to be hereafter published no doubt for the edification of mankind, or womankind." It wasn't sophisticated math, but what Ada said was clear, and complete with admonitions like, "You should never select an *indirect* proof, when a *direct* one can be given." (There's a lot of underlining, here shown as italics, in all Ada's handwritten correspondence.)

Babbage seems at first to have underestimated Ada, trying to interest her in the Silver Lady automaton toy that he used as a conversation piece for his parties (and noting his addition of a turban to it). But Ada continued to interact with (as she put it) Mr. Babbage and Mrs. Somerville, both separately and together. And soon Babbage was opening up to her about many intellectual topics, as well as about the trouble he was having with the government over funding of the Difference Engine.

In the spring of 1835, when Ada was 19, she met 30-year-old William King (or, more accurately, William, Lord King). He was a friend of Mary Somerville's son, had been educated at Eton (the same school where I went 150 years later) and Cambridge, and then had been a civil servant, most recently at an outpost of the British Empire in the Greek islands. William seems to have been a precise, conscientious and decent man, if somewhat stiff. But in any case, Ada and he hit it off, and they were married on July 8, 1835, with Ada keeping the news quiet until the last minute to avoid paparazzi-like coverage.

The next several years of Ada's life seem to have been dominated by having three children and managing a large household—though she had some time for horse riding, learning the harp, and mathematics (including topics like spherical trigonometry). In 1837, Queen Victoria (then 18) came to the throne, and as a member of high society, Ada met her. In 1838, William was made an earl for his government work, and Ada become the Countess of Lovelace.

Within a few months of the birth of her third child in 1839, Ada decided to get more serious about mathematics again. She told Babbage she wanted to find a "mathematical Instructor" in London, though asked that in making enquiries he not mention her name, presumably for fear of society gossip.

The person identified was Augustus De Morgan, first professor of mathematics at University College London, noted logician, author of several textbooks, and not only a friend of Babbage's, but also the husband of the daughter of Ada's mother's main childhood teacher. (Yes, it was a small world. De Morgan was also a friend of George Boole's—and was the person who indirectly caused Boolean algebra to be invented.)

In Ada's correspondence with Babbage, she showed interest in discrete mathematics, and wondered, for example, if the game of solitaire "admits of being put into a mathematical Formula, and solved." But in keeping with the math education traditions of the time (and still today), De Morgan set Ada on studying calculus.

Ockham
16th Feby. 1840

My Dear Mr Babbage.
Have you ever seen a game, or rather puzzle, called Solitaire? — There is an octagonal Board, like the enclosed drawing, with 37 little holes upon it in the position I have drawn them, & 37 little pegs to

Her letters to De Morgan about calculus are not unlike letters from a calculus student today—except for the Victorian English. Even many of the confusions are the same—though Ada was more sensitive than some to the bad notations of calculus ("why can't one multiply by dx?", etc.). Ada was a tenacious student, and seemed to have had a great time learning more and more about mathematics. She was pleased by the mathematical abilities she discovered in herself, and by De Morgan's positive feedback about them. She also continued to interact with

Babbage, and on one visit to her estate (in January 1841, when she was 25), she charmingly told the then-49-year-old Babbage, "If you are a *Skater*, pray bring *Skates* to Ockham; that being the fashionable occupation here now, & one *I* have much taken to."

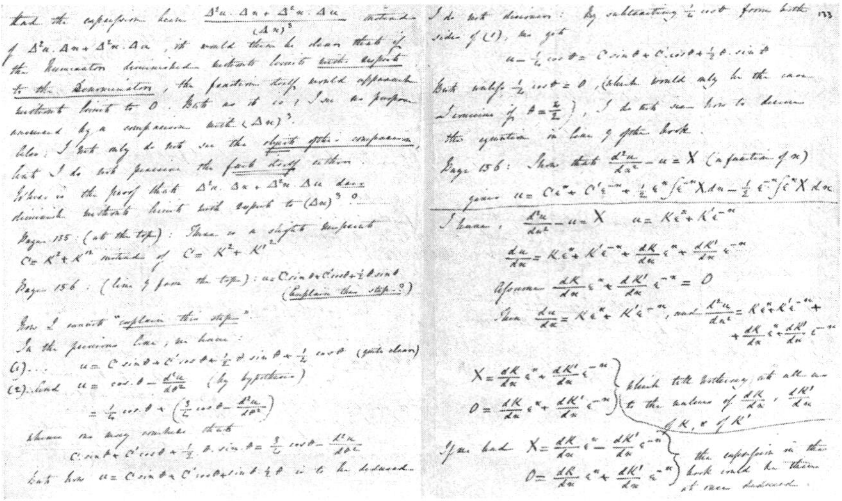

Ada's relationship with her mother was a complex one. Outwardly, Ada treated her mother with great respect. But in many ways she seems to have found her controlling and manipulative. Ada's mother was constantly announcing that she had medical problems and might die imminently (she actually lived to age 64). And she increasingly criticized Ada for her child rearing, household management and deportment in society. But by February 6, 1841, Ada was feeling good enough about herself and her mathematics to write a very open letter to her mother about her thoughts and aspirations.

She wrote, "I believe myself to possess a most singular combination of qualities exactly fitted to make me pre-eminently a discoverer of the hidden realities of nature." She talked of her ambition to do great things. She talked of her "insatiable & restless energy" which she believed she finally had found a purpose for. And she talked about how after 25 years she had become less "secretive & suspicious" with respect to her mother.

But then, three weeks later, her mother dropped a bombshell, claiming that before Ada was born, Byron and his half-sister had had a child together. Incest like that wasn't actually illegal in England at the time, but it was scandalous. Ada took the whole thing very hard, and it derailed her from mathematics.

Ada had had intermittent health problems for years, but in 1841 they apparently worsened, and she started systematically taking opiates. She was very keen to excel in something, and began to get the idea that perhaps it should be music and literature rather than math. But her husband William seems to have talked her out of this, and by late 1842 she was back to doing mathematics.

Back to Babbage

What had Babbage been up to while all this had been going on? He'd been doing all sorts of things, with varying degrees of success.

After several attempts, he'd rather honorifically been appointed Lucasian Professor of Mathematics at Cambridge—but never really even spent time in Cambridge. Still, he wrote what turned out to be a fairly influential book, *On the Economy of Machinery and Manufactures*, dealing with such things as how to break up tasks in factories (an issue that had actually come up in connection with the human computation of mathematical tables).

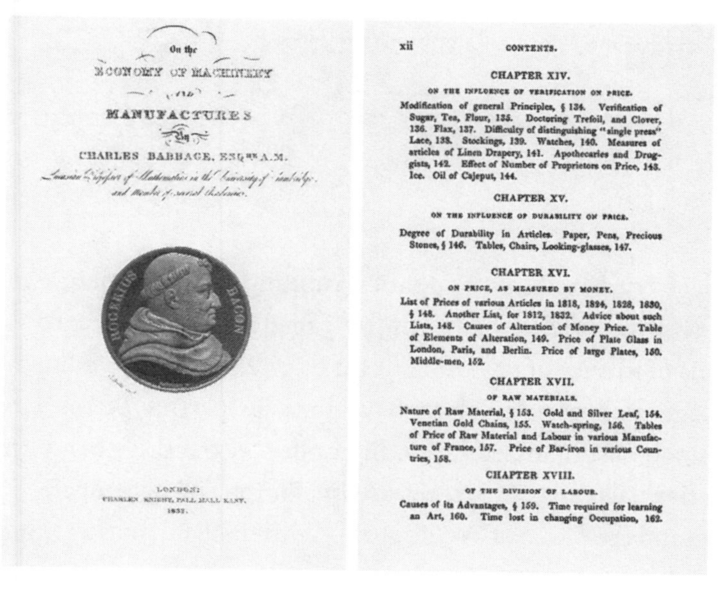

On the
ECONOMY OF MACHINERY
and
MANUFACTURES
By
CHARLES BABBAGE, ESQ.RE A.M.

ROGERIUS BACON

LONDON:
CHARLES KNIGHT, PALL MALL EAST.
1832.

xii CONTENTS.

In 1837, he weighed in on the then-popular subject of natural theology, appending his *Ninth Bridgewater Treatise* to the series of treatises written by other people. The central question was whether there is evidence of a deity from the apparent design seen in nature. Babbage's book is quite hard to read, opening for example with, "The notions we acquire of contrivance and design arise from comparing our observations on the works of other beings with the intentions of which we are conscious in our own undertakings."

In apparent resonance with some of my own work 150 years later, he talks about the relationship between mechanical processes, natural laws and free will. He makes statements like "computations of great complexity can be effected by mechanical means", but then goes on to claim (with rather weak examples) that a mechanical engine can produce sequences of numbers that show unexpected changes that are like miracles.

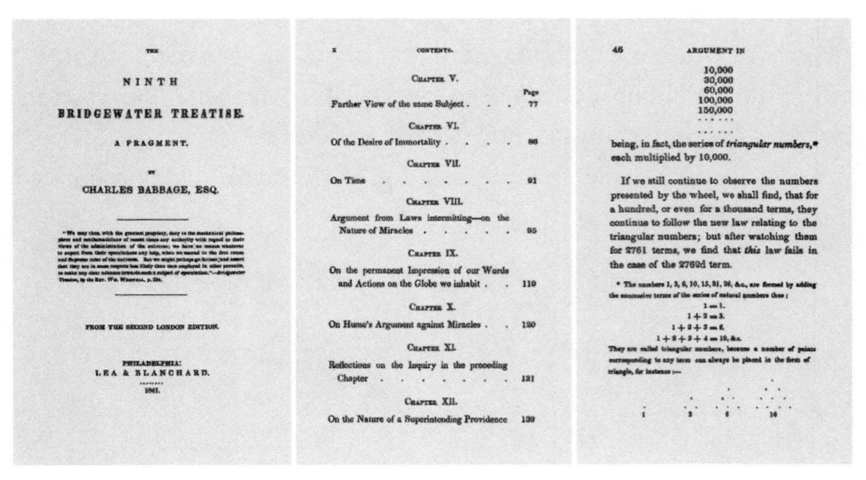
THE

NINTH

BRIDGEWATER TREATISE.

A FRAGMENT.

BY

CHARLES BABBAGE, ESQ.

[illegible]

FROM THE SECOND LONDON EDITION.

PHILADELPHIA:
LEA & BLANCHARD.
1841.

x CONTENTS.

CHAPTER V.

Farther View of the same Subject . . . Page 77

CHAPTER VI.

Of the Desire of Immortality 86

CHAPTER VII.

On Time 91

CHAPTER VIII.

Argument from Laws intermitting—on the Nature of Miracles 95

CHAPTER IX.

On the permanent Impression of our Words and Actions on the Globe we inhabit . . 110

CHAPTER X.

On Hume's Argument against Miracles . . 120

CHAPTER XI.

Reflections on the Inquiry in the preceding Chapter 131

CHAPTER XII.

On the Nature of a Superintending Providence 139

46 ARGUMENT IN

10,000
30,000
60,000
100,000
150,000
......
......

being, in fact, the series of *triangular numbers*,* each multiplied by 10,000.

If we still continue to observe the numbers presented by the wheel, we shall find, that for a hundred, or even for a thousand terms, they continue to follow the new law relating to the triangular numbers; but after watching them for 2761 terms, we find that *this* law fails in the case of the 2762d term.

* The numbers 1, 3, 6, 10, 15, 21, 28, &c., are formed by adding the successive terms of the series of natural numbers thus;

1 = 1.
1 + 2 = 3.
1 + 2 + 3 = 6.
1 + 2 + 3 + 4 = 10, &c.

They are called triangular numbers, because a number of points corresponding to any term can always be placed in the form of triangle, for instance:—

1 3 6 10

Babbage tried his hand at politics, running for parliament twice on a manufacturing-oriented platform, but failed to get elected, partly because of claims of misuse of government funds on the Difference Engine.

Babbage also continued to have upscale parties at his large and increasingly disorganized house in London, attracting such luminaries as Charles Dickens, Charles Darwin, Florence Nightingale, Michael Faraday and the Duke of Wellington—with his aged mother regularly in attendance. But even though the degrees and honors that he listed after his name ran to 6 lines, he was increasingly bitter about his perceived lack of recognition.

CHARLES BABBAGE, ESQ., M.A.,
F.R.S., F.R.S.E., F.R.A.S., F. STAT. S., HON. M.R.I.A., M.C.P.S.,
COMMANDER OF THE ITALIAN ORDER OF ST. MAURICE AND ST. LAZARUS,
INSP. IMP. (ACAD. MORAL.) PARIS CORR., ACAD. AMER. ART. ET SC. BOSTON. REG. ŒCON. BORUSS.,
PHYS. HIST. NAT. GENEV., ACAD. REG. MONAC., HAFN., MASSIL., ET DIVION., SOCIUS.
ACAD. IMP. ET REG. PETROP., NEAP., BRUX., PATAV., GEORG. FLOREN, LYNCEI ROM., MUT., PHILOMATH.
PARIS, SOC. CORR., ETC.

Central to this was what had happened with the Difference Engine. Babbage had hired one of the leading engineers of his day to actually build the engine. But somehow, after a decade of work—and despite lots of precision machine tool development—the actual engine wasn't done. Back in 1833, shortly after he met Ada, Babbage had tried to rein in the project—but the result was that his engineer quit, and insisted that he got to keep all the plans for the Difference Engine, even the ones that Babbage himself had drawn.

But right around this time, Babbage decided he'd had a better idea anyway. Instead of making a machine that would just compute differences, he imagined an "Analytical Engine" that supported a whole list of possible kinds of operations, that could in effect be done in an arbitrarily programmed sequence. At first, he just thought about having the machine evaluate fixed formulas, but as he studied different use cases, he added other capabilities, like conditionals—and figured out often very clever ways to implement them mechanically. But, most important, he figured out how to control the steps in a computation using punched cards of the kind that had been invented in 1801 by Jacquard for specifying patterns of weaving on looms.

List of Operations

1. Algebraic Addition
2. ——— Subtraction
3. Alg.c Add.n with nk figures
4. —— Subt.n with nk figures
5. Ascertaining if a variable is zero
6. ——— has a + or — sign
7. Reversing the Accident.l sign of a variable
8. Stepping up or multiplying any number by $10^{\pm n}$
9. Stepping down or mult.' any number by 10^{n}
10. Count.g No. of digits in Variable & sending it to Store
11. Differences
12. Multiplication without Table.

Babbage created some immensely complicated designs, and today it seems remarkable that they could work. But back in 1826 Babbage had invented something he called Mechanical Notation—that was intended to provide a symbolic representation for the operation of machinery in the same kind of way that mathematical notation provides a symbolic representation for operations in mathematics.

Babbage was disappointed already in 1826 that people didn't appreciate his invention. Undoubtedly people didn't understand it, since even now it's not clear how it worked. But it may have been Babbage's greatest invention—because apparently it's what let him figure out all his elaborate designs.

Babbage's original Difference Engine project had cost the British government £17,500 or the equivalent of perhaps $2 million today. It was a modest sum relative to other government expenditures, but the project was unusual enough to lead to a fair amount of discussion. Babbage was fond of emphasizing that—unlike many of his contemporaries—he hadn't taken government money himself (despite chargebacks for renovating his stable as a fireproof workshop, etc.). He also claimed that he eventually spent £20,000 of his own money—or the majority of his fortune (no, I don't see how the numbers add up)—on his various projects. And he kept on trying to get further government support, and created plans for a Difference Engine No. 2, requiring only 8000 parts instead of 25,000.

By 1842, the government had changed, and Babbage insisted on meeting with the new prime minister (Robert Peel), but ended up just berating him. In parliament the idea of funding the Difference Engine was finally killed with quips like that the machine should be set to compute when it would be of use. (The transcripts of debates about the Difference Engine have a certain charm—especially when they discuss its possible uses for state statistics that strangely parallel computable-country opportunities with Wolfram|Alpha today.)

Ada's Paper

Despite the lack of support in England, Babbage's ideas developed some popularity elsewhere, and in 1840 Babbage was invited to lecture on the Analytical Engine in Turin, and given honors by the Italian government.

Babbage had never published a serious account of the Difference Engine, and had never published anything at all about the Analytical Engine. But he talked about the Analytical Engine in Turin, and notes were taken by a certain Luigi Menabrea, who was then a 30-year-old army engineer—but who, 27 years later, became prime minister of Italy (and also made contributions to the mathematics of structural analysis).

In October 1842, Menabrea published a paper in French based on his notes. When Ada saw the paper, she decided to translate it into English and submit it to a British publication. Many years later Babbage claimed he suggested to Ada that she write her own account of the Analytical Engine, and that she had responded that the thought hadn't occurred to her. But in any case, by February 1843, Ada had resolved to do the translation but add extensive notes of her own.

352 MACHINE ANALYTIQUE

manière que les ondes à la surface des eaux tranquilles et que, dans la formation de ces ondulations, les eaux sont ébranlées aux profondeurs les plus considérables.

Les observations des marées offrent un vif intérêt scientifique; la géologie tirera un grand parti de la connaissance exacte des oscillations de l'Océan; car, avec leur mouvement de transmission plus ou moins rapide, on pourra déterminer avec assez d'exactitude, la profondeur de la mer bien loin du littoral.

NOTIONS SUR LA MACHINE ANALYTIQUE DE M. CHARLES BABBAGE, par Mr. L.-F. MENABREA, capitaine du génie militaire.

Les travaux qui appartiennent à plusieurs branches des sciences mathématiques, quoique paraissant, au premier abord, être uniquement du ressort de l'esprit, peuvent néanmoins se diviser en deux parties distinctes : l'une qu'on peut appeler mécanique, parce qu'elle est sujette à des lois précises et invariables, susceptibles d'être traduites physiquement, tandis que l'autre qui exige l'intervention du raisonnement, est plus spécialement du domaine de la pensée. Dès lors on pourra se proposer de faire exécuter par le moyen de machines la partie mécanique du travail, et réserver à la seule intelligence celle qui dépend de la faculté de raisonner. Ainsi la rigueur à laquelle sont soumises les règles du calcul numérique a dû, depuis longtemps, faire songer à employer des instruments matériels, soit pour exécuter entièrement ces calculs, soit pour les abréger. De là sont nées plusieurs inventions dirigées vers ce but, mais qui ne l'atteignent, en général, qu'imparfaitement. Ainsi la machine de Pascal, tant vantée, n'est maintenant qu'un simple objet de curiosité qui, tout en prouvant une

DE M. CHARLES BABBAGE. 363

NOMBRE des OPÉRATIONS.	CARTONS des OPÉRATIONS. — Signes indiquant la nature des opérations.	CARTONS DES VARIABLES. Colonne soumise aux opérations.	CARTONS DES VARIABLES. Colonne recevant le résultat des opérations	MARCHE des OPÉRATIONS.
1	×	$V_2 \times V_4 =$	V_8	$= dn'$
2	×	$V_5 \times V_1 =$	V_9	$= d'n$
3	×	$V_4 \times V_0 =$	V_{10}	$= n'm$
4	×	$V_1 \times V_3 =$	V_{11}	$= nm'$
5	—	$V_8 - V_9 =$	V_{12}	$= dn' - d'n$
6	—	$V_{10} - V_{11} =$	V_{13}	$= n'm - n'm'$
7	:	$\frac{V_{12}}{V_{13}} =$	V_{14}	$= x = \frac{dn' - d'n}{nm' - m'n}$

Comme les cartons ne font qu'indiquer comment et sur quelles colonnes la machine doit agir, il est clair qu'il faudra encore, dans chaque cas particulier, introduire les données numériques du calcul. Ainsi, dans l'exemple que nous avons choisi, on devra préalablement écrire les valeurs numériques de m, n, d, m', n', d' dans l'ordre et sur les colonnes indiquées, après quoi l'on fera agir la machine qui donnera la valeur de l'inconnue x pour ce cas particulier. Pour avoir la valeur de y, il faudra faire une autre série d'opérations analogues aux précédentes. Mais on voit qu'elles se réduiront à quatre seulement, car le dénominateur de l'expression de y, sauf le signe, est le même que celui de x, et égal à $m'n - mn'$. Dans le tableau précédent l'on remarquera que la colonne des opérations indique de suite quatre *multiplications*, deux *soustractions* et une *division*. Ainsi l'on pourra, au besoin, n'employer que trois cartons des opérations; pour cela il suffira d'introduire dans la machine un appareil qui, par exemple, après la première multiplication, retienne le carton relatif à cette opération, et ne lui permette d'avancer, pour être remplacé par

Over the months that followed she worked very hard—often exchanging letters almost daily with Babbage (despite sometimes having other "pressing and unavoidable engagements"). And though in those days letters were sent by post (which did come 6 times a day in London at the time) or carried by a servant (Ada lived about a mile from Babbage when she was in London), they read a lot like emails about a project might today, apart from being in Victorian English. Ada asks Babbage questions; he responds; she figures things out; he comments on them. She was clearly in charge, but felt she was first and foremost explaining Babbage's work, so wanted to check things with him—though she got annoyed when Babbage, for example, tried to make his own corrections to her manuscript.

It's charming to read Ada's letter as she works on debugging her computation of Bernoulli numbers: "My Dear Babbage. I am in much dismay at having got into so amazing a quagmire & botheration with these *Numbers*, that I cannot possibly get the thing done today. .... I am now going out on horseback. Tant mieux." Later she told Babbage, "I have worked incessantly, & most successfully, all day. You will admire the Table & Diagram extremely. They have been made out with extreme care, & all the indices most minutely & scrupulously attended to." Then she added that William (or "Lord L." as she referred to him) "is at this moment kindly inking it all over for me. I had to do it in pencil..."

July 1843 382
Friday. 4 o'clock
21. July?

My dear Babbage. I am in much dismay at having got into so amazing a quagmire & botheration with these *Numbers*, that I cannot possibly get the thing done today. I have no doubt it will all come out clear enough tomorrow; & I shall send you a *packet* up, as early in the day as I can. So do not be uneasy. (Tho' at this moment I am in a charming state of confusion; but it is that sort of confusion which is of a very *bubble* nature.)

I am now going out on horseback. Tant mieux.

Yours puzzle-pate

William was also apparently the one who suggested that she sign the translation and notes. As she wrote to Babbage, "It is not my wish to *proclaim* who has written it; at the same time I rather wish to append anything that may tend hereafter to *individualize*, & *identify* it, with the other productions of the said A.A.L." (for "Ada Augusta Lovelace").

By the end of July 1843, Ada had pretty much finished writing her notes. She was proud of them, and Babbage was complimentary about them. But Babbage wanted one more thing: he wanted to add an anonymous preface (written by him) that explained how the British government had failed to support the project. Ada thought it a bad idea. Babbage tried to insist, even suggesting that without the preface the whole publication should be withdrawn. Ada was furious, and told Babbage so. In the end, Ada's translation appeared, signed "AAL", without the preface, followed by her notes headed "Translator's Note".

Ada was clearly excited about it, sending reprints to her mother, and explaining that "No one can estimate the trouble & *interminable* labour of having to revise the printing of *mathematical* formulae. This is a pleasant prospect for the future, as I suppose many hundreds & thousands of such formulae will come forth from my pen, in one way or another." She said that her husband William had been excitedly giving away copies to his friends too, and Ada wrote, "William especially conceives that it places me in a much *juster* & *truer* position & light, than anything else can. And he tells me that it has already placed *him* in a far more agreeable position in this country."

Within days, there was also apparently society gossip about Ada's publication. She explained to her mother that she and William "are by no means desirous of making it a secret, altho' I do not wish the *importance* of the thing to be exaggerated and overrated". She saw herself as being a successful expositor and interpreter of Babbage's work, setting it in a broader conceptual framework that she hoped could be built on.

There's lots to say about the actual content of Ada's notes. But before we get to that, let's finish the story of Ada herself.

While Babbage's preface wasn't itself a great idea, one good thing it did for posterity was to cause Ada on August 14, 1843 to write Babbage a fascinating, and very forthright, 16-page letter. (Unlike her usual letters, which were on little folded pages, this was on large sheets.) In it, she explains that while he is often "implicit" in what he says, she is herself "always a very 'explicit function of x'". She says that "Your affairs have been, & are, deeply occupying both myself and Lord Lovelace.... And the result is that I have plans for you..." Then she proceeds to ask, "If I am to lay before you in the course of a year or two, explicit & honorable propositions for *executing your engine* ... would there be any chance of allowing myself ... to conduct the business for you; your own *undivided* energies being devoted to the execution of the work..."

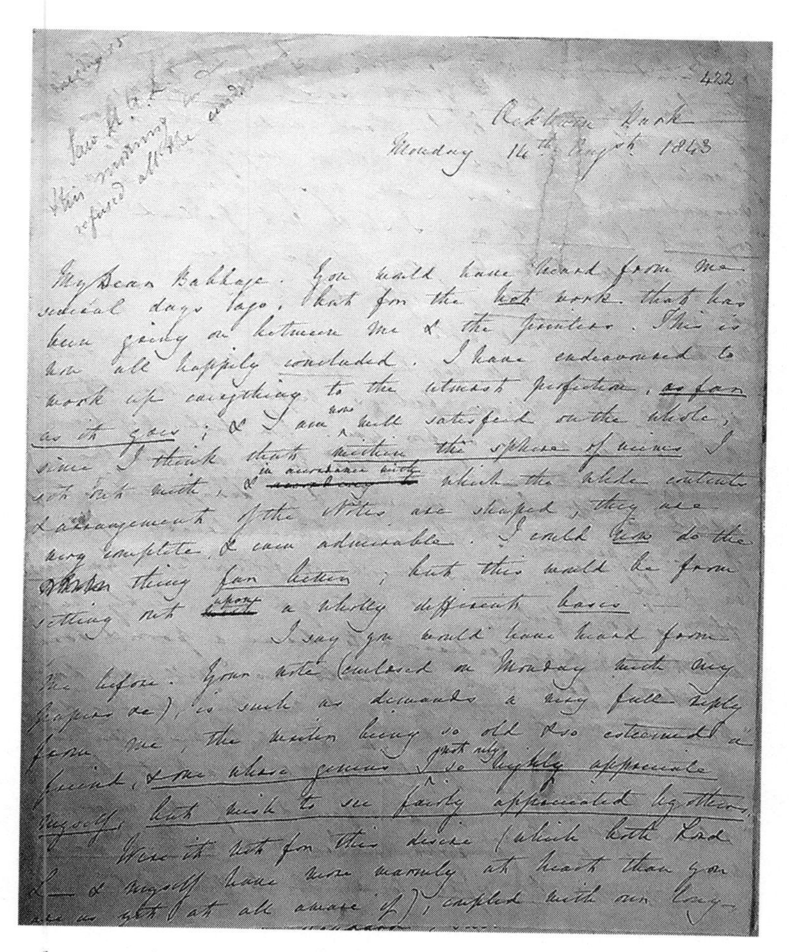

422

Ockham Park
Monday 14th August 1843

My Dear Babbage. You would have heard from me several days ago, but for the hot work that has been going on between me & the printers. This is now all happily concluded. I have endeavoured to work up everything to the utmost perfection, as far as it goes; & I am now well satisfied on the whole; since I think that within the sphere of views I set out with, & in accordance with which the whole contents & arrangement of the Notes are shaped, they are very complete & even admirable. I could now do the same thing far better; but this would be from setting out upon a wholly different basis.

I say you would have heard from me before. Your note (enclosed on Monday with my papers &c), is such as demands a very full reply from me, the writer being so old & so esteemed a friend, & one whose genius I so highly appreciate myself, but wish to see fairly appreciated by others.

Were it not for this desire (which both Lord L— & myself have more warmly at heart than you are as yet at all aware of), coupled with our long-

In other words, she basically proposed to take on the role of CEO, with Babbage becoming CTO. It wasn't an easy pitch to make, especially given Babbage's personality. But she was skillful in making her case, and as part of it, she discussed their different motivation structures. She wrote, "My own uncompromising principle is to endeavour to love *truth & God before fame & glory* ...", while "Yours is to love truth & God ... but to love *fame, glory, honours, yet more.*" Still, she explained, "Far be it from me, to disclaim the influence of ambition & fame. No living soul ever was more imbued with it than myself ... but I certainly would not deceive myself or others by pretending it is other than a very important motive & ingredient in my character & nature."

She ended the letter, "I wonder if you will choose to retain the lady-fairy in your service or not."

At noon the next day she wrote to Babbage again, asking if he would help in "the *final* revision". Then she added, "You will have had my long letter this morning. Perhaps you will not choose to have anything more to do with me. But I hope the best..."

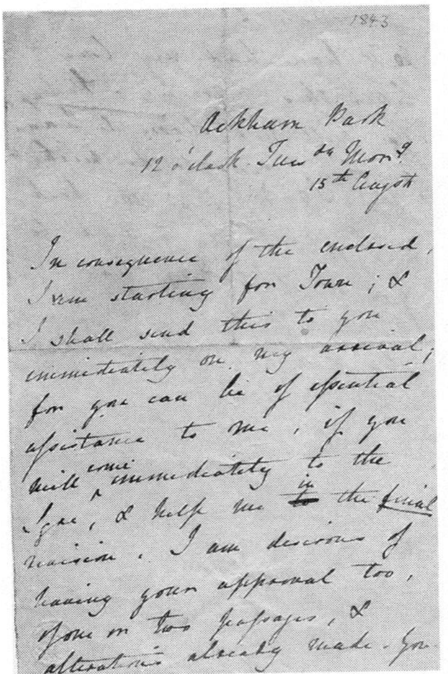

1843

Ockham Park
12 o'clock Tuesday Morng
15th August

In consequence of the enclosed, I am starting for Town; & I shall send this to you immediately on my arrival; for you can be of essential assistance to me, if you will come immediately to the Square, & help me in the final revision. I am desirous of having your approval too, of one or two passages, & alterations already made. You will have had my long letter this morning. Perhaps you will not choose to have anything more to do with me. But I hope the best, & so I just write & send to you as if nothing had passed.

You will have this at half after 2 o'clock.

A. L

At 5 pm that day, Ada was in London, and wrote to her mother, "I am uncertain as yet how the Babbage business will end.... I have written to him ... very explicitly; stating my own *conditions* ... He has so strong an idea of the *advantage* of having *my* pen as his servant, that he will probably yield; though I demand very strong concessions. If he *does* consent to what I propose, I shall probably be enabled to keep him out of much hot water; & to bring his engine to *consummation*, (which all I have seen of him & his habits the last 3 months, makes me scarcely anticipate it ever *will* be, unless someone really exercises a strong coercive influence over him). He is beyond measure *careless* & *desultory* at times. — I shall be willing to be his Whipper-in during the next 3 years if I see fair prospect of success."

But on Babbage's copy of Ada's letter, he scribbled, "Saw A.A.L. this morning and refused all the conditions."

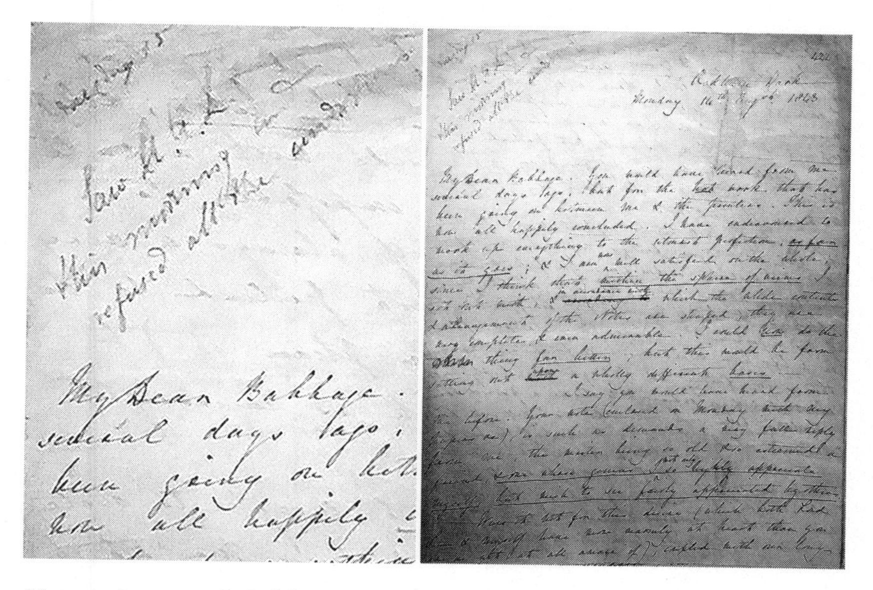
Saw A.A.L. this morning and refused all the conditions

Monday 16th August 1848

My Dear Babbage. You would have learned from me several days ago, that from the work that has been going on between me & the printers. This is now all happily concluded.

Yet on August 18, Babbage wrote to Ada about bringing drawings and papers when he would next come to visit her. The next week, Ada wrote to Babbage that "We are quite delighted at your (somewhat *unhoped* for) proposal" [of a long visit with Ada and her husband]. And Ada wrote to her mother, "Babbage & I are I think more friends than ever. I have never seen him so agreeable, so reasonable, or in such good spirits!"

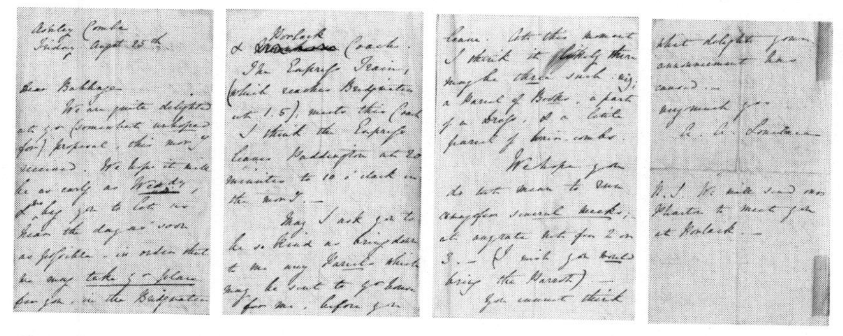
Ashley Combe
Friday August 25th

Dear Babbage
We are quite delighted at your (somewhat unhoped for) proposal, this morning received. We hope it will be as early as Wednesday, & beg you to let us hear the day as soon as possible — in order that we may take your place for you in the Bridgewater & Porlock Coach. The Express Train, (which reaches Bridgwater at 1.5), meets this Coach. I think the Express leaves Paddington at 20 minutes to 10 o'clock in the morning. —

May I ask you to be so kind as bring down to me any Parcels which may be sent to your house for me, before you leave. At this moment I think it probably there may be three such, viz, a Parcel of Books, a part of a Dress, & a little parcel of hair-combs.

We hope you do not mean to run away for several weeks, at any rate not for 2 or 3. — (I wish you would bring the Harass) —

You cannot think what delight your announcement has caused. —

Very much yours
A. A. Lovelace

P.S. We will send our Phaeton to meet you at Porlock. —

Then, on September 9, Babbage wrote to Ada, expressing his admiration for her and (famously) describing her as "Enchantress of Number" and "my dear and much admired Interpreter". (Yes, despite what's often quoted, he wrote "Number" not "Numbers".)

My dear Lady Lovelace
I find it quite in vain to wait until I have leisure so I have resolved that I will leave all other things undone and set out for Ashley taking with me papers enough to enable me to forget ~~this~~ world and all its troubles and if possible its multitudinous Charlatans—every thing in short but the Enchantress of Numbers.

The next day, Ada responded to Babbage, "You are a brave man to give yourself wholly up to Fairy-Guidance!" and Babbage signed off on his next letter as "Your faithful Slave." And Ada described herself to her mother as serving as the "High-Priestess of Babbage's Engine."

After the Paper

But unfortunately that's not how things worked out. For a while it was just that Ada had to take care of household and family things that she'd neglected while concentrating on her Notes. But then her health collapsed, and she spent many months going between doctors and various "cures" (her mother suggested "mesmerism", i.e. hypnosis), all the while watching their effects on, as she put it, "that portion of the material forces of the world entitled the body of A.A.L."

She was still excited about science, though. She'd interacted with Michael Faraday, who apparently referred to her as "the *rising star* of Science". She talked about her first publication as her "first-born", "with a colouring & undercurrent (rather *hinted at* & *suggested* than definitely expressed) of *large, general, & metaphysical views*", and said that "He [the publication] will make an excellent head (I hope) of a large family of brothers & sisters".

When her notes were published, Babbage had said, "You should have written an original paper. The postponement of that will however only render it more perfect." But by October 1844, it seemed that David Brewster (inventor of the kaleidoscope, among other things) would write about the Analytical Engine, and Ada asked if perhaps Brewster could suggest another topic for her, saying, "I rather think some physiological topics would suit me as well as any."

And indeed later that year, she wrote to a friend (who was also her lawyer, as well as being Mary Somerville's son), "It does not appear to me that cerebral matter need be more unmanageable to mathematicians than *sidereal* & *planetary* matter & movements; if they would but inspect it from the *right point of view*. I hope to bequeath to the generations a *Calculus of the Nervous System*." An impressive vision—10 years before, for example, George Boole would talk about similar things.

Both Babbage and Mary Somerville had started their scientific publishing careers with translations, and she saw herself as doing the same, saying that perhaps her next works would be reviews of Whewell and Ohm, and that she might eventually become a general "prophet of science".

There were roadblocks, to be sure. Like that, at that time, as a woman, she couldn't get access to the Royal Society's library in London, even though her husband, partly through her efforts, was a member of the society. But the most serious issue was still Ada's health. She had a whole series of problems, though in 1846 she was still saying optimistically, "Nothing is needed but a year or two more of patience & *cure*."

There were also problems with money. William had a never-ending series of elaborate—and often quite innovative—construction projects (he seems to have been particularly keen on towers and tunnels). And to finance them, they had to turn to Ada's mother, who often made things difficult. Ada's children were also approaching teenage-hood, and Ada was exercised by many issues that were coming up with them.

Meanwhile, she continued to have a good social relationship with Babbage, seeing him with considerable frequency, though in her letters talking more about dogs and pet parrots than the Analytical Engine. In

1848 Babbage developed a hare-brained scheme to construct an engine that played tic-tac-toe, and to tour it around the country as a way to raise money for his projects. Ada talked him out of it. The idea was raised for Babbage to meet Prince Albert to discuss his engines, but it never happened.

William also dipped his toe into publishing. He had already written short reports with titles like "Method of growing Beans and Cabbages on the same Ground" and "On the Culture of Mangold-Wurzel". But in 1848 he wrote one more substantial piece, comparing the productivity of agriculture in France and England, based on detailed statistics, with observations like, "It is demonstrable, not only that the Frenchman is much worse off than the Englishman, but that he is less well fed than during the devastating exhaustion of the empire."

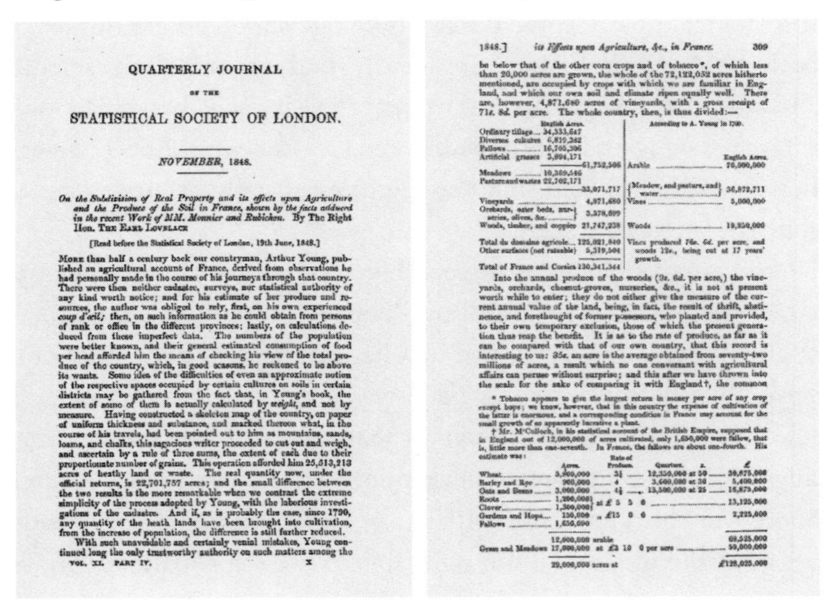

QUARTERLY JOURNAL

OF THE

STATISTICAL SOCIETY OF LONDON.

NOVEMBER, 1848.

On the Subdivision of Real Property and its effects upon Agriculture and the Produce of the Soil in France, shown by the facts adduced in the recent Work of MM. Monnier and Rubichon. By The Right Hon. THE EARL LOVELACE

[Read before the Statistical Society of London, 19th June, 1848.]

MORE than half a century back our countryman, Arthur Young, published an agricultural account of France, derived from observations he had personally made in the course of his journeys through that country. There were then neither cadastre, surveys, nor statistical authority of any kind worth notice; and for his estimate of her produce and resources, the author was obliged to rely, first, on his own experienced *coup d'œil;* then, on such information as he could obtain from persons of rank or office in the different provinces; lastly, on calculations deduced from these imperfect data. The numbers of the population were better known, and their general estimated consumption of food per head afforded him the means of checking his view of the total produce of the country, which, in good seasons, he reckoned to be above its wants. Some idea of the difficulties of even an approximate notion of the respective spaces occupied by certain cultures on soils in certain districts may be gathered from the fact that, in Young's book, the extent of some of them is actually calculated by *weight*, and not by measure. Having constructed a skeleton map of the country, on paper of uniform thickness and substance, and marked thereon what, in the course of his travels, had been pointed out to him as mountains, sands, loams, and chalks, this sagacious writer proceeded to cut out and weigh, and ascertain by a rule of three sums, the extent of each due to their proportionate number of grains. This operation afforded him 25,513,213 acres of heathy land or waste. The real quantity now, under the official returns, is 22,701,757 acres; and the small difference between the two results is the more remarkable when we contrast the extreme simplicity of the process adopted by Young, with the laborious investigations of the cadastre. And if, as is probably the case, since 1790, any quantity of the heath lands have been brought into cultivation, from the increase of population, the difference is still further reduced.

With such unavoidable and certainly venial mistakes, Young continued long the only trustworthy authority on such matters among the

VOL. XI. PART IV. X

1848.] *its Effects upon Agriculture, &c., in France.* 309

be below that of the other corn crops and of tobacco*, of which less than 20,000 acres are grown, the whole of the 72,122,052 acres hitherto mentioned, are occupied by crops with which we are familiar in England, and which our own soil and climate ripen equally well. There are, however, 4,871,680 acres of vineyards, with a gross receipt of 71*s.* 8*d.* per acre. The whole country, then, is thus divided:—

	English Acres.		According to A. Young in 1790.	English Acres.
Ordinary tillage	34,333,647			
Diverses cultures	6,819,382			
Fallows	16,705,306			
Artificial grasses	3,894,171			
		61,752,506	Arable	70,000,000
Meadows	10,369,546			
Pasture and wastes	22,702,171			
		33,071,717	Meadow, and pasture, and water	36,872,711
Vineyards		4,871,680	Vines	5,000,000
Orchards, osier beds, nurseries, olives, &c.		3,578,699		
Woods, timber, and coppice		21,747,238	Woods	19,850,000
Total du domaine agricole		125,021,840	Vines produced 76*s.* 6*d.* per acre, and woods 12*s.*, being cut at 17 years' growth.	
Other surfaces (not rateable)		5,319,504		
Total of France and Corsica		130,341,344		

Into the annual produce of the woods (9*s.* 6*d.* per acre,) the vineyards, orchards, chesnut-groves, nurseries, &c., it is not at present worth while to enter; they do not either give the measure of the current annual value of the land, being, in fact, the result of thrift, abstinence, and forethought of former possessors, who planted and provided, to their own temporary exclusion, those of which the present generation thus reap the benefit. It is as to the rate of produce, as far as it can be compared with that of our own country, that this record is interesting to us: 35*s.* an acre is the average obtained from seventy-two millions of acres, a result which no one conversant with agricultural affairs can peruse without surprise; and this after we have thrown into the scale for the sake of comparing it with England†, the common

* Tobacco appears to give the largest return in money per acre of *any crop* except hops; we know, however, that in this country the expense of cultivation of the latter is enormous, and a corresponding condition in France may account for the small growth of so apparently lucrative a plant.

† Mr. M'Culloch, in his statistical account of the British Empire, supposed that in England out of 12,000,000 of acres cultivated, only 1,650,000 were fallow, that is, little more than one-seventh. In France, the fallows are about one-fourth. His estimate was:

	Acres.	Rate of Produce.	Quarters.	s.	£
Wheat	3,800,000	3½	12,350,000	at 50	30,875,000
Barley and Rye	900,000	4	3,600,000	at 30	5,400,000
Oats and Beans	3,000,000	4½	13,500,000	at 25	16,875,000
Roots	1,200,000	at £5 5 0			13,125,000
Clover	1,300,000				
Gardens and Hops	150,000	„ £15 0 0			2,250,000
Fallows	1,650,000				
	12,000,000 arable				68,525,000
Grass and Meadows	17,000,000	at £3 10 0 per acre			59,500,000
	29,000,000 acres at				£128,025,000

1850 was a notable year for Ada. She and William moved into a new house in London, intensifying their exposure to the London scientific social scene. She had a highly emotional experience visiting for the first time her father's family's former estate in the north of England—and got into an argument with her mother about it. And she got more deeply involved in betting on horseracing, and lost some money doing it. (It wouldn't have been out of character for Babbage or her to invent some mathematical scheme for betting, but there's no evidence that they did.)

In May 1851 the Great Exhibition opened at the Crystal Palace in London. (When Ada visited the site back in January, Babbage advised, "Pray put on worsted stockings, cork soles and every other thing which can keep you warm.") The exhibition was a high point of Victorian science and technology, and Ada, Babbage and their scientific social circle were all involved (though Babbage less so than he thought he should be). Babbage gave out many copies of a flyer on his Mechanical Notation. William won an award for brick-making.

But within a year, Ada's health situation was dire. For a while her doctors were just telling her to spend more time at the seaside. But eventually they admitted she had cancer (from what we know now, probably cervical cancer). Opium no longer controlled her pain; she experimented with cannabis. By August 1852, she wrote, "I begin to understand Death; which is going on quietly & gradually every minute, & will never be a thing of one particular moment." And on August 19, she asked Babbage's friend Charles Dickens to visit and read her an account of death from one of his books.

Her mother moved into her house, keeping other people away from her, and on September 1, Ada made an unknown confession that apparently upset William. She seemed close to death, but she hung on, in great pain, for nearly 3 more months, finally dying on November 27, 1852, at the age of 36. Florence Nightingale, nursing pioneer and friend of Ada's, wrote, "They said she could not possibly have lived so long, were it not for the tremendous vitality of the brain, that would not die."

Ada had made Babbage the executor of her will. And—much to her mother's chagrin—she had herself buried in the Byron family vault next to her father, who, like her, died at age 36 (Ada lived 266 days longer). Her mother built a memorial that included a sonnet entitled "The Rainbow" that Ada wrote.

Inscribed by the express direction of
ADA AUGUSTA LOVELACE,
Born December 10th 1816, died November 27th 1852,
To recall her memory.

And the prayer of faith shall save the sick,
And the Lord shall raise him up;
And if he have committed sins,
They shall be forgiven him.

Bow down in hope, in thanks all ye that mourn
Where'er this peerless arch of radiant hues,
Surpassing earthly tints, the storm subdues
Of Nature's smiles and tears. 'Tis heaven-born
To soothe the sad, the sinning, and forlorn;
A lovely loving token to inspire
The hope, the faith, that Power divine endues
With latent good the woes by which we're torn.
'Tis like a sweet repentance of the skies,
To beckon all by sense of sin opprest,
Revealing harmony from sin and sighs;
A pledge that deep implanted in the breast
A hidden light may burn that never dies
And bursts through storms in purest hues exprest.

The Aftermath

Ada's funeral was small; neither her mother nor Babbage attended. But the obituaries were kind, if Victorian in their sentiments:

LADY LOVELACE.

WHO has not felt an interest in the only child of Byron, the Ada whose name is so caressed in his verse, and a lock of whose hair is the subject of a touching passage in his letters? Who has not felt at least a curiosity to know what features of genius and character had descended from the father to the daughter? The Countess of Lovelace was thoroughly original, and the poetic temperament was all that was hers in common with her father. Her genius, for genius she possessed, was not poetic, but metaphysical and mathematical, her mind having been in the constant practice of investigation, and this with rigorous exactness. With an understanding thoroughly masculine in solidity, grasp, and firmness, Lady Lovelace had all the delicacies of the most refined female character. Her manners, her tastes, her accomplishments, in many of which, Music especially, she was a proficient, were feminine in the nicest sense of the word; and the superficial observer would never have divined the strength and the knowledge that lay hidden under the womanly graces. Proportionate to her distaste for the frivolous and commonplace was her enjoyment of true intellectual society, and eagerly she sought the acquaintance of all who were distinguished in science, art, and literature. But from this pleasure, and all else, in the prime of life she has been cut off. She bore a long and painful illness with the fortitude, the heroism belonging to her character. We need not add to this feeble, imperfect tribute how deeply she must be mourned by all honoured with her friendship—a friendship so cordial, so frank.

William outlived her by 41 years, eventually remarrying. Her oldest son—with whom Ada had many difficulties—joined the navy several years before she died, but deserted. Ada thought he might have gone to America (he was apparently in San Francisco in 1851), but in fact he died at 26 working in a shipyard in England. Ada's daughter married a somewhat wild poet, spent many years in the Middle East, and became the world's foremost breeder of Arabian horses. Ada's youngest son inherited the family title, and spent most of his life on the family estate.

Ada's mother died in 1860, but even then the gossip about her and Byron continued, with books and articles appearing, including Harriet Beecher Stowe's 1870 *Lady Byron Vindicated*. In 1905, a year before he died, Ada's youngest son—who had been largely brought up by Ada's mother—published a book about the whole thing, with such choice lines as "Lord Byron's life contained nothing of any interest except what ought not to have been told".

When Ada died, there was a certain air of scandal that seemed to hang around her. Had she had affairs? Had she run up huge gambling debts? There's scant evidence of either. Perhaps it was a reflection of her father's "bad boy" image. But before long there were claims that she'd pawned the family jewels (twice!), or lost, some said, £20,000, or maybe even £40,000 (equivalent to about $7 million today) betting on horses.

It didn't help that Ada's mother and her youngest son both seemed against her. On September 1, 1852—the same day as her confession to William—Ada had written, "It is my earnest and dying request that all my friends who have letters from me will deliver them to my mother Lady Noel Byron after my death." Babbage refused. But others complied, and, later on, when her son organized them, he destroyed some.

But many thousands of pages of Ada's documents still remain, scattered around the world. Back-and-forth letters that read like a modern text stream, setting up meetings, or mentioning colds and other ailments. Charles Babbage complaining about the postal service. Three Greek sisters seeking money from Ada because their dead brother had been a page for Lord Byron. Charles Dickens talking about chamomile tea. Pleasantries from a person Ada met at Paddington Station. And household accounts, with entries for note paper, musicians, and ginger biscuits. And then, mixed in with all the rest, serious intellectual discussion about the Analytical Engine and many other things.

What Happened to Babbage?

So what happened to Babbage? He lived 18 more years after Ada, dying in 1871. He tried working on the Analytical Engine again in 1856, but made no great progress. He wrote papers with titles like "On the Statistics of Light-Houses", "Table of the Relative Frequency of Occurrences of the Causes of Breaking Plate-Glass Windows", and "On Remains of Human Art, mixed with the Bones of Extinct Races of Animals".

Then in 1864 he published his autobiography, *Passages from the Life of a Philosopher*—a strange and rather bitter document. The chapter on the Analytical Engine opens with a quote from a poem by Byron—"Man wrongs, and Time avenges"—and goes on from there. There are chapters on "Theatrical experience", "Hints for travellers" (including on advice about how to get an RV-like carriage in Europe), and, perhaps most peculiar, "Street nuisances". For some reason Babbage waged a campaign against street musicians who he claimed woke him up at 6 am, and caused him to lose a quarter of his productive time. One wonders why he didn't invent a sound-masking solution, but his campaign was so notable, and so odd, that when he died it was a leading element of his obituary.

Babbage never remarried after his wife died, and his last years seem to have been lonely ones. A gossip column of the time records impressions of him:

196 A TWILIGHT GOSSIP WITH THE PAST.

played on till long after midnight. Babbage was a first-rate player."

From Sir Andrew I heard the story, which I believe has been told elsewhere, of Whewell, Peacock, and Babbage walking together across the quadrangle of Trinity, when Peacock observed, "Will, I think we can boast that we are the three ugliest fellows in the University." "Speak for yourself, Mr. Peacock," retorted Whewell in evident annoyance, and turning round, left his friends to the consideration of "how in minutiæ the character peeps out."

Babbage was a plain man, I must allow, the plainest of the three, I think, but he wore well; in the quarter of a century that I knew him he had scarcely altered at all. Early in the sixties Miss Kinglake and I went one evening to take tea with Mr. Babbage. He had promised to show us some interesting papers respecting Lady Lovelace's mathematical studies, and by arrangement there were no other guests. Mr. Babbage's house in Dorset Street, Manchester Square, was the same that had long been occupied by Dr. Wollaston. It was large and rambling for a London house, having several spacious sitting-rooms, all of which, with the exception of the drawing-room, were crammed with books, papers, and apparatus in apparent confusion, but the philosopher knew where to put his hand on everything. He received us in his unused drawing-room, which looked dreary in the extreme; the furniture had the stiff primness of age and pretension without a trace of homely use and custom. No one could have turned the cat out of the most comfortable chair as Chaucer's monk did, for no such chair was ever there. The place was dimly lighted by four candles, the grate yawned black and fireless, for it was not yet winter. Coleridge said he did not believe in ghosts, he had seen too many. But no bold sense of scepticism relieved me from the creepy feeling of that hour; every chair had its ghost, and I fancied thin, disembodied forms crowding in at the further end of the room. Was it a trick that the glass gave back a reflection, that was not myself, though I alone stood in front? I am inclined to think that that bewildering and hateful function known as "a spring cleaning" might have rectified the false impressions of the ghostly mirror; but this was an after-thought.

I do not remember in my whole life a more curious and noteworthy evening than the one I am now describing. Mr. Babbage had reached his *anecdotage*, was in the mood to be communicative, and my friend Miss Kinglake, claiming the privilege of age, asked without reserve such questions as induced

A TWILIGHT GOSSIP WITH THE PAST. 197

our host to be autobiographical. He told us that not only had he crippled his private fortune by his devotion to his calculating machine, but that for this idol of his brain he had given up all the pleasures and comforts of domestic life. He married early, but his wife died while he was a young man. With an amount of feeling that I had never associated with a philosopher who wore the armour of cynicism, he pathetically lamented the dreary isolation of his lot, "for of course," said he, "fond as I am of domestic life, I should have married again if it had not been for my machine."

Mr. Babbage was always reticent about his early life, so much so that an impression got about that he was of humble birth. This was not the case. He was born in 1792 at Totnes, in a good old mansion in the town; twenty years ago it was known as "The Castle Inn," and perhaps is still so called. His father, a well-to-do banker, was nicknamed by his townsfolk "Old Five per cents," he talked so constantly about money matters. The eminent mathematician's mother lived to a great age, and I have heard from those who remembered earlier times, that she was occasionally to be seen at the brilliant receptions Babbage used to give in the forties, in this same dreary, ghost-haunted room where we then sat, with its faded hangings and tarnished gilding. In the old days it was the son's greatest pleasure to bring up his most distinguished guests to be introduced to his mother, the homely old lady seated on the stiff-backed sofa, the place of honour.

He spoke of his mother on this memorable evening, repeating to us her reply when it became a question whether he should make further outlay respecting the machine, which had already cost his private purse £20,000. The old lady said with a large-mindedness rare in our thrifty sex, "My dear son, you have a great object in view worthy of your ambition; my advice is, pursue it, even if it should oblige you to live on bread and cheese." Babbage mentions the fact, I believe, in his 'Passages in the Life of a Philosopher.' I well remember Sir Andrew Ramsay praising this book very highly, as being an autobiography of permanent interest, adding that the world often owes more to the impetus given to progress by a man's mind than to his completed work. This was said of course in reference to the calculating machine, which seems to me to have been the bane of his life. I speak as a non-mathematician, and am therefore unworthy to speak; but with Babbage's great powers and practical capacity, his country would gladly have associated his name with something other than a magnificent failure. His conversation on

Apparently he was fond of saying that he would gladly give up the remainder of his life if he could spend just 3 days 500 years in the future. When he died, his brain was preserved, and is still on display...

Even though Babbage never finished his Difference Engine, a Swedish company did, and even already displayed part of it at the Great Exhibition. When Babbage died, many documents and spare parts from his Difference Engine project passed to his son Major-General Henry Babbage, who published some of the documents, and privately assembled a few more devices, including part of the Mill for the Analytical Engine. Meanwhile, the fragment of the Difference Engine that had been built in Babbage's time was deposited at the Science Museum in London.

Rediscovery

After Babbage died, his life work on his engines was all but forgotten (though he did, for example, get a mention in the 1911 *Encyclopædia Britannica*). Mechanical computers nevertheless continued to be developed, gradually giving way to electromechanical ones, and eventually to electronic ones. And when programming began to be understood in the 1940s, Babbage's work—and Ada's Notes—were rediscovered.

People knew that "AAL" was Ada Augusta Lovelace, and that she was Byron's daughter. Alan Turing read her Notes, and coined the term "Lady Lovelace's Objection" ("an AI can't originate anything") in his 1950 Turing Test paper. But Ada herself was still largely a footnote at that point.

It was a certain Bertram Bowden—a British nuclear physicist who went into the computer industry and eventually became Minister of Science and Education—who "rediscovered" Ada. In researching his 1953 book *Faster Than Thought* (yes, about computers), he located Ada's granddaughter Lady Wentworth (the daughter of Ada's daughter), who told him the family lore about Ada, both accurate and inaccurate, and let him look at some of Ada's papers. Charmingly, Bowden notes that in Ada's granddaughter's book *Thoroughbred Racing Stock*, there is use of binary in computing pedigrees. Ada, and the Analytical Engine, of course, used decimal, with no binary in sight.

But even in the 1960s, Babbage—and Ada—weren't exactly well known. Babbage's Difference Engine prototype had been given to the Science Museum in London, but even though I spent lots of time at the Science Museum as a child in the 1960s, I'm pretty sure I never saw it there. Still, by the 1980s, particularly after the US Department of Defense named its ill-fated programming language after Ada, awareness of Ada Lovelace and Charles Babbage began to increase, and biographies began to appear, though sometimes with hair-raising errors (my favorite is that the mention of "the problem of three bodies" in a letter from Babbage indicated a romantic triangle between Babbage, Ada and William—while it actually refers to the three-body problem in celestial mechanics!).

As interest in Babbage and Ada increased, so did curiosity about whether the Difference Engine would actually have worked if it had been built from Babbage's plans. A project was mounted, and in 2002, after a heroic effort, a complete Difference Engine was built, with only one correction in the plans being made. Amazingly, the machine worked. Building it cost about the same, inflation adjusted, as Babbage had requested from the British government back in 1823.

What about the Analytical Engine? So far, no real version of it has ever been built—or even fully simulated.

What Ada Actually Wrote

OK, so now that I've talked (at length) about the life of Ada Lovelace, what about the actual content of her Notes on the Analytical Engine?

They start crisply: "The particular function whose integral the Difference Engine was constructed to tabulate, is...." She then explains that the Difference Engine can compute values of any 6th degree polynomial—but the Analytical Engine is different, because it can perform any sequence of operations. Or, as she says, "The Analytical Engine is an *embodying of the science of operations*, constructed with peculiar reference to abstract number as the subject of those operations. The Difference Engine is the embodying of one particular and very limited set of operations..."

Charmingly, at least for me, considering the years I have spent working on Mathematica, she continues at a later point, "We may consider the engine as the *material and mechanical representative of analysis*, and that our actual working powers in this department of human study will be enabled more effectually than heretofore to keep pace with our theoretical knowledge of its principles and laws, through the complete control which the engine gives us over the executive manipulation of algebraical and numerical symbols."

A little later, she explains that punched cards are how the Analytical Engine is controlled, and then makes the classic statement that "the Analytical Engine *weaves algebraical patterns* just as the Jacquard-loom weaves flowers and leaves".

Ada then goes through how a sequence of specific kinds of computations would work on the Analytical Engine, with "Operation Cards" defining the operations to be done, and "Variable Cards" defining the locations of values. Ada talks about "cycles" and "cycles of cycles, etc", now known as loops and nested loops, giving a mathematical notation for them:

(6.) $(\div), \Sigma(+1)^p (\times, -)$ or $(1), \Sigma(+1)^p (2, 3)$,

where p stands for the variable; $(+1)^p$ for the function of the variable, that is, for ϕp; and the limits are from 1 to p, or from 0 to $p-1$, each increment being equal to unity. Similarly, (4.) would be,—

(7.) $$\Sigma(+1)^n \left\{(\div), \Sigma(+1)^p (\times, -)\right\}$$

the limits of n being from 1 to n, or from 0 to $n-1$,

(8.) $$\text{or } \Sigma(+1)^n \left\{(1), \Sigma(+1)^p (2, 3)\right\}.$$

There's a lot of modern-seeming content in Ada's notes. She comments that "There is in existence a beautiful woven portrait of Jacquard, in the fabrication of which 24,000 cards were required." Then she discusses the idea of using loops to reduce the number of cards needed, and the value of rearranging operations to optimize their execution on the Analytical Engine, ultimately showing that just 3 cards could do what might seem like it should require 330.

Ada talks about just how far the Analytical Engine can go in computing what was previously not computable, at least with any accuracy. And as an example she discusses the three-body problem, and the fact that in her time, of "about 295 coefficients of lunar perturbations" there were many on which different people's computations didn't agree.

Finally comes Ada's Note G. Early on, she states, "The Analytical Engine has no pretensions whatever to *originate* anything. It can do whatever we *know how to order it* to perform. ... Its province is to assist us in making available what we are already acquainted with."

Ada seems to have understood with some clarity the traditional view of programming: that we engineer programs to do things we know how to do. But she also notes that in actually putting "the truths and the formulae of analysis" into a form amenable to the engine, "the nature of many subjects in that science are necessarily thrown into new lights, and more profoundly investigated." In other words—as I often point out—actually programming something inevitably lets one do more exploration of it.

She goes on to say that "in devising for mathematical truths a new form in which to record and throw themselves out for actual use, views are likely to be induced, which should again react on the more theoretical phase of the subject", or in other words—as I have also often said—representing mathematical truths in a computable form is likely to help one understand those truths themselves better.

Ada seems to have understood, though, that the "science of operations" implemented by the engine would not only apply to traditional mathematical operations. For example, she notes that if "the fundamental relations of pitched sounds in the science of harmony" were amenable to abstract operations, then the engine could use them to "compose elaborate and scientific pieces of music of any degree of complexity or extent." Not a bad level of understanding for 1843.

The Bernoulli Number Computation

What's become the most famous part of what Ada wrote is the computation of Bernoulli numbers, in Note G. This seems to have come out of a letter she wrote to Babbage, in July 1843. She begins the letter with, "I am working very hard for you; like the Devil in fact; (which perhaps I am)". Then she asks for some specific references, and finally ends with, "I want to put in something about Bernoulli's Numbers, in one of my Notes, as an example of how an implicit function may be worked out by the engine, without having been worked out by human head & hands first.... Give me the necessary data & formulae."

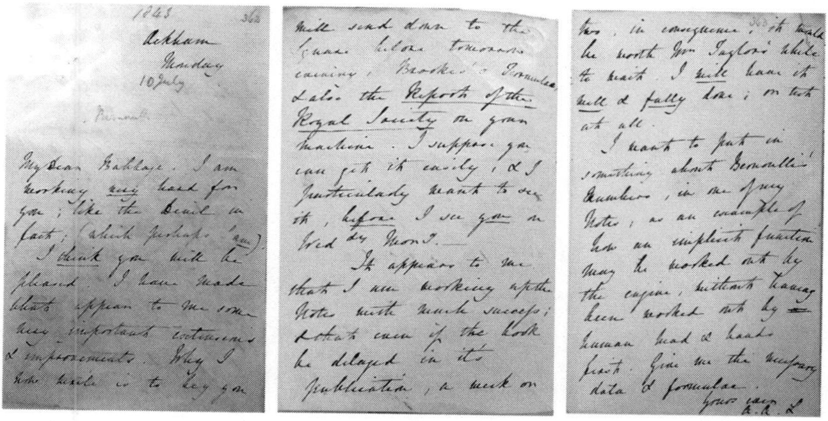

1843
Ockham
Monday
10 July

My Dear Babbage. I am working very hard for you; like the Devil in fact; (which perhaps I am).
I think you will be pleased. I have made what appear to me some very important extensions & improvements. Why I now write is to beg you will send down to the Square before tomorrow evening, Brooke's Formulae, & also the Reports of the Royal Society on your machine. I suppose you can get it easily; & I particularly want to see it, before I see you on Wed[y] Mon[y].—
It appears to me that I am working up the Notes with much success; & that even if the book be delayed in it's publication, a week or two, in consequence, it would be worth Mr Taylor's while to wait. I will have it well & fully done; or not at all.
I want to put in something about Bernoulli's Numbers, in one of my Notes, as an example of how an implicit function may be worked out by the engine, without having been worked out by human head & hands first. Give me the necessary data & formulae.
Yours ever
A.A.L

Ada's choice of Bernoulli numbers to show off the Analytical Engine was an interesting one. Back in the 1600s, people spent their lives making tables of sums of powers of integers—in other words, tabulating values of $\sum_{k=1}^{m} k^n$ for different m and n. But Jacob Bernoulli pointed out that all such sums can be expressed as polynomials in m, with the coefficients being related to what are now called Bernoulli numbers. And in 1713 Bernoulli was proud to say that he'd computed the first 10 Bernoulli numbers "in a quarter of an hour"—reproducing years of other people's work.

Today, of course, it's instantaneous to do the computation in the Wolfram Language:

```
In[1]:= Table[BernoulliB[n], {n, 0, 10}]
```

$$\text{Out[1]=} \left\{1, -\frac{1}{2}, \frac{1}{6}, 0, -\frac{1}{30}, 0, \frac{1}{42}, 0, -\frac{1}{30}, 0, \frac{5}{66}\right\}$$

And, as it happens, a few years ago, just to show off new algorithms, we even computed 10 million of them.

But, OK, so how did Ada plan to do it? She started from the fact that Bernoulli numbers appear in the series expansion:

$$\frac{x}{e^x - 1} = \sum_{n=0}^{\infty} \frac{B_n\, x^n}{n!}$$

Then by rearranging this and matching up powers of x, she got a sequence of equations for the Bernoulli numbers B_n—which she then "unravelled" to give a recurrence relation of the form:

$$B_n = \frac{1}{n+1} \sum_{k=0}^{n-1} \binom{n+1}{k} B_k$$

Now Ada had to specify how to actually compute this on the Analytical Engine. First, she used the fact that odd Bernoulli numbers (other than B_1) are zero, then computed B_n, which is our modern B_{2n} (or BernoulliB[2n] in Wolfram Language). Then she started from B_0, and successively computed B_n for larger n, storing each value she got. The algorithm she used for the computation was (in modern terms):

```
B[0]=1
B[n_]:=
  B[n]=1/2 (2 n - 1)/(2 n + 1) - Sum[B[j] Product[i, {i, 2 n - 2 j + 2, 2 n}]/Product[i, {i, 2 j}], {j, n - 1}]
```

On the Analytical Engine, the idea was to have a sequence of operations (specified by "Operation Cards") performed by the "Mill" with operands coming from the "Store" (with addresses specified by "Variable Cards"). (In the Store, each number was represented by a sequence of wheels, each turned to the appropriate value for each digit.) To compute Bernoulli numbers the way Ada wanted takes two nested loops of operations. With the Analytical Engine design that existed at the time, Ada had to basically unroll these loops. But in the end she successfully produced a description of how B_8 (which she called B_7) could be computed:

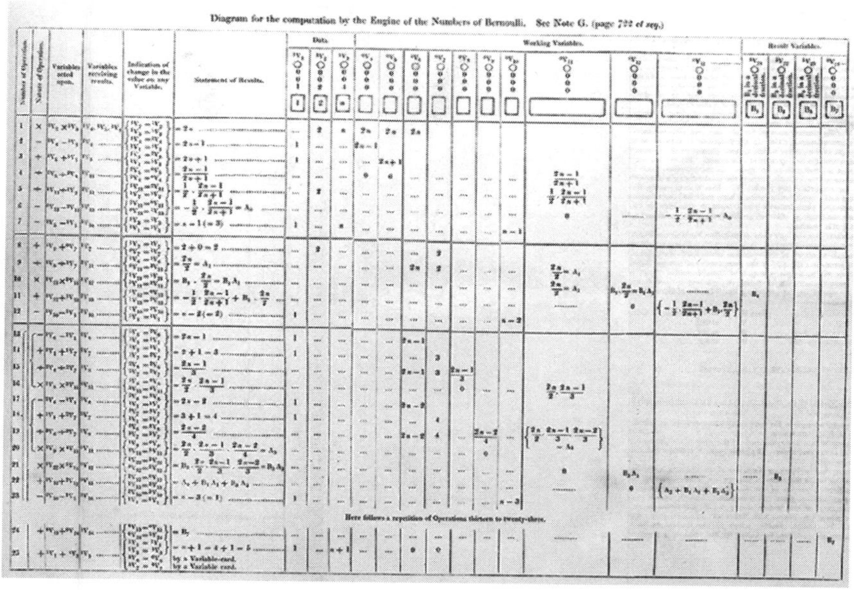

Diagram for the computation by the Engine of the Numbers of Bernoulli. See Note G. (page 722 et seq.)

Here follows a repetition of Operations thirteen to twenty-three.

This is effectively the execution trace of a program that runs for 25 steps (plus a loop) on the Analytical Engine. At each step, the trace shows what operation is performed on which Variable Cards, and which Variable Cards receive the results. Lacking a symbolic notation for loops, Ada just indicated loops in the execution trace using braces, noting in English that parts are repeated.

And in the end, the final result of the computation appears in location 24:

In[2]:= **BernoulliB[8]**

Out[2]= $-\frac{1}{30}$

As it's printed, there's a bug in Ada's execution trace on line 4: the fraction is upside down. But if you fix that, it's easy to get a modern version of what Ada did:

	Operation	Data			Working Variables										Results			
		V_1	V_2	V_3	V_4	V_5	V_6	V_7	V_8	V_9	V_{10}	V_{11}	V_{12}	V_{13}	V_{21}	V_{22}	V_{23}	V_{24}
		1	2	4	0	0	0	0	0	0	0	0	0	0	B_1	B_3	B_5	B_7
1	$V_4=V_5=V_6=V_2\times V_3$	1	2	4	8	8	8	0	0	0	0	0	0	0				
2	$V_4=V_4-V_1$	1	2	4	7	8	8	0	0	0	0	0	0	0				
3	$V_5=V_5+V_1$	1	2	4	7	9	8	0	0	0	0	0	0	0				
4	$V_{11}=V_4/V_5$	1	2	4	0	9	8	0	0	0	0	$\frac{7}{9}$	0	0				
5	$V_{11}=V_{11}/V_2$	1	2	4	0	9	8	0	0	0	0	$\frac{7}{18}$	0	0				
6	$V_{13}=V_{13}-V_{11}$	1	2	4	0	9	8	0	0	0	0	0	0	$-\frac{7}{18}$				
7	$V_{10}=V_3-V_1$	1	2	4	0	9	8	0	0	0	3	0	0	$-\frac{7}{18}$				
8	$V_7=V_2-V_7$	1	2	4	0	9	8	2	0	0	3	0	0	$-\frac{7}{18}$				
9	$V_{11}=V_6/V_7$	1	2	4	0	9	8	2	0	0	3	4	0	$-\frac{7}{18}$				
10	$V_{12}=V_{21}\times V_{11}$	1	2	4	0	9	8	2	0	0	3	4	$\frac{2}{3}$	$-\frac{7}{18}$	$\frac{1}{6}$			
11	$V_{13}=V_{12}+V_{13}$	1	2	4	0	9	8	2	0	0	3	4	0	$\frac{5}{18}$				
12	$V_{10}=V_{10}-V_1$	1	2	4	0	9	8	2	0	0	2	4	0	$\frac{5}{18}$				
13	$V_6=V_6-V_1$	1	2	4	0	9	7	2	0	0	2	4	0	$\frac{5}{18}$				
14	$V_7=V_1+V_7$	1	2	4	0	9	7	3	0	0	2	4	0	$\frac{5}{18}$				
15	$V_8=V_6/V_7$	1	2	4	0	9	7	3	$\frac{7}{3}$	0	2	4	0	$\frac{5}{18}$				
16	$V_{11}=V_8\times V_{11}$	1	2	4	0	9	7	3	0	0	2	$\frac{28}{3}$	0	$\frac{5}{18}$				
17	$V_6=V_6-V_1$	1	2	4	0	9	6	3	0	0	2	$\frac{28}{3}$	0	$\frac{5}{18}$				
18	$V_7=V_1+V_7$	1	2	4	0	9	6	4	0	0	2	$\frac{28}{3}$	0	$\frac{5}{18}$				
19	$V_9=V_6/V_7$	1	2	4	0	9	6	4	0	$\frac{3}{2}$	2	$\frac{28}{3}$	0	$\frac{5}{18}$				
20	$V_{11}=V_9\times V_{11}$	1	2	4	0	9	6	4	0	0	2	14	0	$\frac{5}{18}$				
21	$V_{12}=V_{22}\times V_{11}$	1	2	4	0	9	6	4	0	0	2	14	$-\frac{7}{15}$	$\frac{5}{18}$		$-\frac{1}{30}$		
22	$V_{13}=V_{12}+V_{13}$	1	2	4	0	9	6	4	0	0	2	14	0	$-\frac{17}{90}$				
23	$V_{10}=V_{10}-V_1$	1	2	4	0	9	6	4	0	0	1	14	0	$-\frac{17}{90}$				
24	$V_6=V_6-V_1$	1	2	4	0	9	5	4	0	0	1	14	0	$-\frac{17}{90}$				
25	$V_7=V_1+V_7$	1	2	4	0	9	5	5	0	0	1	14	0	$-\frac{17}{90}$				
26	$V_8=V_6/V_7$	1	2	4	0	9	5	5	1	0	1	14	0	$-\frac{17}{90}$				
27	$V_{11}=V_8\times V_{11}$	1	2	4	0	9	5	5	0	0	1	14	0	$-\frac{17}{90}$				
28	$V_6=V_6-V_1$	1	2	4	0	9	4	5	0	0	1	14	0	$-\frac{17}{90}$				
29	$V_7=V_1+V_7$	1	2	4	0	9	4	6	0	0	1	14	0	$-\frac{17}{90}$				
30	$V_9=V_6/V_7$	1	2	4	0	9	4	6	0	$\frac{2}{3}$	1	14	0	$-\frac{17}{90}$				
31	$V_{11}=V_9\times V_{11}$	1	2	4	0	9	4	6	0	0	1	$\frac{28}{3}$	0	$-\frac{17}{90}$				
32	$V_{12}=V_{23}\times V_{11}$	1	2	4	0	9	4	6	0	0	1	$\frac{28}{3}$	$\frac{2}{9}$	$-\frac{17}{90}$			$\frac{1}{42}$	
33	$V_{13}=V_{12}+V_{13}$	1	2	4	0	9	4	6	0	0	1	$\frac{28}{3}$	0	$\frac{1}{30}$				
34	$V_{10}=V_{10}-V_1$	1	2	4	0	9	4	6	0	0	0	$\frac{28}{3}$	0	$\frac{1}{30}$				
35	$V_{24}=V_{13}+V_{24}$	1	2	4	0	9	4	6	0	0	0	$\frac{28}{3}$	0	$\frac{1}{30}$				$-\frac{1}{30}$
36	$V_3=V_1+V_3$	1	2	5	0	9	0	0	0	0	0	$\frac{28}{3}$	0	$\frac{1}{30}$				

And here's what the same scheme gives for the next two (nonzero) Bernoulli numbers. As Ada figured out it doesn't ultimately take any more working variables (specified by Variable Cards) to compute higher Bernoulli numbers, just more operations.

	Operation	Data			Working Variables										Results				
		V_1	V_2	V_3	V_4	V_5	V_6	V_7	V_8	V_9	V_{10}	V_{11}	V_{12}	V_{13}	V_{21}	V_{22}	V_{23}	V_{24}	V_{25}
		1	2	5	0	0	0	0	0	0	0	0	0	0	B_1	B_3	B_5	B_7	B_9
1	$V_4=V_5=V_6=V_2\times V_3$	1	2	5	10	10	10	0	0	0	0	0	0	0					
2	$V_4=V_4-V_1$	1	2	5	9	10	10	0	0	0	0	0	0	0					
3	$V_5=V_5+V_1$	1	2	5	9	11	10	0	0	0	0	0	0	0					
4	$V_{11}=V_4/V_5$	1	2	5	0	11	10	0	0	0	0	$\frac{9}{11}$	0	0					
5	$V_{11}=V_{11}/V_2$	1	2	5	0	11	10	0	0	0	0	$\frac{9}{22}$	0	0					
6	$V_{13}=V_{13}-V_{11}$	1	2	5	0	11	10	0	0	0	0	0	0	$-\frac{9}{22}$					
7	$V_{10}=V_3-V_1$	1	2	5	0	11	10	0	0	0	4	0	0	$-\frac{9}{22}$					
8	$V_7=V_2-V_7$	1	2	5	0	11	10	2	0	0	4	0	0	$-\frac{9}{22}$					
9	$V_{11}=V_6/V_7$	1	2	5	0	11	10	2	0	0	4	5	0	$-\frac{9}{22}$					
10	$V_{12}=V_{21}\times V_{11}$	1	2	5	0	11	10	2	0	0	4	5	$\frac{5}{6}$	$-\frac{9}{22}$	$\frac{1}{6}$				
11	$V_{13}=V_{12}+V_{13}$	1	2	5	0	11	10	2	0	0	4	5	0	$\frac{14}{33}$					
12	$V_{10}=V_{10}-V_1$	1	2	5	0	11	10	2	0	0	3	5	0	$\frac{14}{33}$					
13	$V_6=V_6-V_1$	1	2	5	0	11	9	2	0	0	3	5	0	$\frac{14}{33}$					
14	$V_7=V_1+V_7$	1	2	5	0	11	9	3	0	0	3	5	0	$\frac{14}{33}$					
15	$V_8=V_6/V_7$	1	2	5	0	11	9	3	3	0	3	5	0	$\frac{14}{33}$					
16	$V_{11}=V_8\times V_{11}$	1	2	5	0	11	9	3	0	0	3	15	0	$\frac{14}{33}$					
17	$V_6=V_6-V_1$	1	2	5	0	11	8	3	0	0	3	15	0	$\frac{14}{33}$					
18	$V_7=V_1+V_7$	1	2	5	0	11	8	4	0	0	3	15	0	$\frac{14}{33}$					
19	$V_9=V_6/V_7$	1	2	5	0	11	8	4	0	2	3	15	0	$\frac{14}{33}$					
20	$V_{11}=V_9\times V_{11}$	1	2	5	0	11	8	4	0	0	3	30	0	$\frac{14}{33}$					
21	$V_{12}=V_{22}\times V_{11}$	1	2	5	0	11	8	4	0	0	3	30	-1	$\frac{14}{33}$		$-\frac{1}{30}$			
22	$V_{13}=V_{12}+V_{13}$	1	2	5	0	11	8	4	0	0	3	30	0	$-\frac{19}{33}$					
23	$V_{10}=V_{10}-V_1$	1	2	5	0	11	8	4	0	0	2	30	0	$-\frac{19}{33}$					
24	$V_6=V_6-V_1$	1	2	5	0	11	7	4	0	0	2	30	0	$-\frac{19}{33}$					
25	$V_7=V_1+V_7$	1	2	5	0	11	7	5	0	0	2	30	0	$-\frac{19}{33}$					
26	$V_8=V_6/V_7$	1	2	5	0	11	7	5	$\frac{7}{5}$	0	2	30	0	$-\frac{19}{33}$					
27	$V_{11}=V_8\times V_{11}$	1	2	5	0	11	7	5	0	0	2	42	0	$-\frac{19}{33}$					
28	$V_6=V_6-V_1$	1	2	5	0	11	6	5	0	0	2	42	0	$-\frac{19}{33}$					
29	$V_7=V_1+V_7$	1	2	5	0	11	6	6	0	0	2	42	0	$-\frac{19}{33}$					
30	$V_9=V_6/V_7$	1	2	5	0	11	6	6	0	1	2	42	0	$-\frac{19}{33}$					
31	$V_{11}=V_9\times V_{11}$	1	2	5	0	11	6	6	0	0	2	42	0	$-\frac{19}{33}$					
32	$V_{12}=V_{23}\times V_{11}$	1	2	5	0	11	6	6	0	0	2	42	1	$-\frac{19}{33}$			$\frac{1}{42}$		
33	$V_{13}=V_{12}+V_{13}$	1	2	5	0	11	6	6	0	0	2	42	0	$\frac{14}{33}$					
34	$V_{10}=V_{10}-V_1$	1	2	5	0	11	6	6	0	0	1	42	0	$\frac{14}{33}$					
35	$V_6=V_6-V_1$	1	2	5	0	11	5	6	0	0	1	42	0	$\frac{14}{33}$					
36	$V_7=V_1+V_7$	1	2	5	0	11	5	7	0	0	1	42	0	$\frac{14}{33}$					
37	$V_8=V_6/V_7$	1	2	5	0	11	5	7	$\frac{5}{7}$	0	1	42	0	$\frac{14}{33}$					
38	$V_{11}=V_8\times V_{11}$	1	2	5	0	11	5	7	0	0	1	30	0	$\frac{14}{33}$					
39	$V_6=V_6-V_1$	1	2	5	0	11	4	7	0	0	1	30	0	$\frac{14}{33}$					
40	$V_7=V_1+V_7$	1	2	5	0	11	4	8	0	0	1	30	0	$\frac{14}{33}$					
41	$V_9=V_6/V_7$	1	2	5	0	11	4	8	0	$\frac{1}{2}$	1	30	0	$\frac{14}{33}$					
42	$V_{11}=V_9\times V_{11}$	1	2	5	0	11	4	8	0	0	1	15	0	$\frac{14}{33}$					
43	$V_{12}=V_{24}\times V_{11}$	1	2	5	0	11	4	8	0	0	1	15	$-\frac{1}{2}$	$\frac{14}{33}$				$-\frac{1}{30}$	
44	$V_{13}=V_{12}+V_{13}$	1	2	5	0	11	4	8	0	0	1	15	0	$-\frac{5}{66}$					
45	$V_{10}=V_{10}-V_1$	1	2	5	0	11	4	8	0	0	0	15	0	$-\frac{5}{66}$					
46	$V_{25}=V_{13}+V_{25}$	1	2	5	0	11	4	8	0	0	0	15	0	$-\frac{5}{66}$					$\frac{5}{66}$
47	$V_3=V_1+V_3$	1	2	6	0	11	0	0	0	0	0	15	0	$-\frac{5}{66}$					

	Operation	Data			Working Variables										Results					
		V_1	V_2	V_3	V_4	V_5	V_6	V_7	V_8	V_9	V_{10}	V_{11}	V_{12}	V_{13}	V_{21}	V_{22}	V_{23}	V_{24}	V_{25}	V_{26}
		1	2	6	0	0	0	0	0	0	0	0	0	0	B_1	B_3	B_5	B_7	B_9	B_{11}
1	$V_4=V_5=V_6=V_2\times V_3$	1	2	6	12	12	12	0	0	0	0	0	0	0						
2	$V_4=V_4-V_1$	1	2	6	11	12	12	0	0	0	0	0	0	0						
3	$V_5=V_5+V_1$	1	2	6	11	13	12	0	0	0	0	0	0	0						
4	$V_{11}=V_4/V_5$	1	2	6	0	13	12	0	0	0	0	$\frac{11}{13}$	0	0						
5	$V_{11}=V_{11}/V_2$	1	2	6	0	13	12	0	0	0	0	$\frac{11}{26}$	0	0						
6	$V_{13}=V_{13}-V_{11}$	1	2	6	0	13	12	0	0	0	0	0	0	$-\frac{11}{26}$						
7	$V_{10}=V_3-V_1$	1	2	6	0	13	12	0	0	0	5	0	0	$-\frac{11}{26}$						
8	$V_7=V_2-V_7$	1	2	6	0	13	12	2	0	0	5	0	0	$-\frac{11}{26}$						
9	$V_{11}=V_6/V_7$	1	2	6	0	13	12	2	0	0	5	6	0	$-\frac{11}{26}$						
10	$V_{12}=V_{21}\times V_{11}$	1	2	6	0	13	12	2	0	0	5	6	1	$-\frac{11}{26}$	$\frac{1}{6}$					
11	$V_{13}=V_{12}+V_{13}$	1	2	6	0	13	12	2	0	0	5	6	0	$\frac{15}{26}$						
12	$V_{10}=V_{10}-V_1$	1	2	6	0	13	12	2	0	0	4	6	0	$\frac{15}{26}$						
13	$V_6=V_6-V_1$	1	2	6	0	13	11	2	0	0	4	6	0	$\frac{15}{26}$						
14	$V_7=V_1+V_7$	1	2	6	0	13	11	3	0	0	4	6	0	$\frac{15}{26}$						
15	$V_8=V_6/V_7$	1	2	6	0	13	11	3	$\frac{11}{3}$	0	4	6	0	$\frac{15}{26}$						
16	$V_{11}=V_8\times V_{11}$	1	2	6	0	13	11	3	0	0	4	22	0	$\frac{15}{26}$						
17	$V_6=V_6-V_1$	1	2	6	0	13	10	3	0	0	4	22	0	$\frac{15}{26}$						
18	$V_7=V_1+V_7$	1	2	6	0	13	10	4	0	0	4	22	0	$\frac{15}{26}$						
19	$V_9=V_6/V_7$	1	2	6	0	13	10	4	0	$\frac{5}{2}$	4	22	0	$\frac{15}{26}$						
20	$V_{11}=V_9\times V_{11}$	1	2	6	0	13	10	4	0	0	4	55	0	$\frac{15}{26}$						
21	$V_{12}=V_{22}\times V_{11}$	1	2	6	0	13	10	4	0	0	4	55	$-\frac{11}{6}$	$\frac{15}{26}$		$-\frac{1}{30}$				
22	$V_{13}=V_{12}+V_{13}$	1	2	6	0	13	10	4	0	0	4	55	0	$-\frac{49}{39}$						
23	$V_{10}=V_{10}-V_1$	1	2	6	0	13	10	4	0	0	3	55	0	$-\frac{49}{39}$						
24	$V_6=V_6-V_1$	1	2	6	0	13	9	4	0	0	3	55	0	$-\frac{49}{39}$						
25	$V_7=V_1+V_7$	1	2	6	0	13	9	5	0	0	3	55	0	$-\frac{49}{39}$						
26	$V_8=V_6/V_7$	1	2	6	0	13	9	5	$\frac{9}{5}$	0	3	55	0	$-\frac{49}{39}$						
27	$V_{11}=V_8\times V_{11}$	1	2	6	0	13	9	5	0	0	3	99	0	$-\frac{49}{39}$						
28	$V_6=V_6-V_1$	1	2	6	0	13	8	5	0	0	3	99	0	$-\frac{49}{39}$						
29	$V_7=V_1+V_7$	1	2	6	0	13	8	6	0	0	3	99	0	$-\frac{49}{39}$						
30	$V_9=V_6/V_7$	1	2	6	0	13	8	6	0	$\frac{4}{3}$	3	99	0	$-\frac{49}{39}$						
31	$V_{11}=V_9\times V_{11}$	1	2	6	0	13	8	6	0	0	3	132	0	$-\frac{49}{39}$						
32	$V_{12}=V_{23}\times V_{11}$	1	2	6	0	13	8	6	0	0	3	132	$\frac{22}{7}$	$-\frac{49}{39}$			$\frac{1}{42}$			
33	$V_{13}=V_{12}+V_{13}$	1	2	6	0	13	8	6	0	0	3	132	0	$\frac{515}{273}$						
34	$V_{10}=V_{10}-V_1$	1	2	6	0	13	8	6	0	0	2	132	0	$\frac{515}{273}$						
35	$V_6=V_6-V_1$	1	2	6	0	13	7	6	0	0	2	132	0	$\frac{515}{273}$						
36	$V_7=V_1+V_7$	1	2	6	0	13	7	7	0	0	2	132	0	$\frac{515}{273}$						
37	$V_8=V_6/V_7$	1	2	6	0	13	7	7	1	0	2	132	0	$\frac{515}{273}$						
38	$V_{11}=V_8\times V_{11}$	1	2	6	0	13	7	7	0	0	2	132	0	$\frac{515}{273}$						
39	$V_6=V_6-V_1$	1	2	6	0	13	6	7	0	0	2	132	0	$\frac{515}{273}$						
40	$V_7=V_1+V_7$	1	2	6	0	13	6	8	0	0	2	132	0	$\frac{515}{273}$						
41	$V_9=V_6/V_7$	1	2	6	0	13	6	8	0	$\frac{3}{4}$	2	132	0	$\frac{515}{273}$						
42	$V_{11}=V_9\times V_{11}$	1	2	6	0	13	6	8	0	0	2	99	0	$\frac{515}{273}$						
43	$V_{12}=V_{24}\times V_{11}$	1	2	6	0	13	6	8	0	0	2	99	$-\frac{33}{10}$	$\frac{515}{273}$				$-\frac{1}{30}$		
44	$V_{13}=V_{12}+V_{13}$	1	2	6	0	13	6	8	0	0	2	99	0	$-\frac{3859}{2730}$						
45	$V_{10}=V_{10}-V_1$	1	2	6	0	13	6	8	0	0	1	99	0	$-\frac{3859}{2730}$						
46	$V_6=V_6-V_1$	1	2	6	0	13	5	8	0	0	1	99	0	$-\frac{3859}{2730}$						
47	$V_7=V_1+V_7$	1	2	6	0	13	5	9	0	0	1	99	0	$-\frac{3859}{2730}$						
48	$V_8=V_6/V_7$	1	2	6	0	13	5	9	$\frac{5}{9}$	0	1	99	0	$-\frac{3859}{2730}$						
49	$V_{11}=V_8\times V_{11}$	1	2	6	0	13	5	9	0	0	1	55	0	$-\frac{3859}{2730}$						
50	$V_6=V_6-V_1$	1	2	6	0	13	4	9	0	0	1	55	0	$-\frac{3859}{2730}$						
51	$V_7=V_1+V_7$	1	2	6	0	13	4	10	0	0	1	55	0	$-\frac{3859}{2730}$						
52	$V_9=V_6/V_7$	1	2	6	0	13	4	10	0	$\frac{2}{5}$	1	55	0	$-\frac{3859}{2730}$						
53	$V_{11}=V_9\times V_{11}$	1	2	6	0	13	4	10	0	0	1	22	0	$-\frac{3859}{2730}$						
54	$V_{12}=V_{25}\times V_{11}$	1	2	6	0	13	4	10	0	0	1	22	$\frac{5}{3}$	$-\frac{3859}{2730}$					$\frac{5}{66}$	
55	$V_{13}=V_{12}+V_{13}$	1	2	6	0	13	4	10	0	0	1	22	0	$\frac{691}{2730}$						
56	$V_{10}=V_{10}-V_1$	1	2	6	0	13	4	10	0	0	0	22	0	$\frac{691}{2730}$						
57	$V_{26}=V_{13}+V_{26}$	1	2	6	0	13	4	10	0	0	0	22	0	$\frac{691}{2730}$						$-\frac{691}{2730}$
58	$V_3=V_1+V_3$	1	2	7	0	13	0	0	0	0	0	22	0	$\frac{691}{2730}$						

The Analytical Engine, as it was designed in 1843, was supposed to store 1000 40-digit numbers, which would in principle have allowed it to compute up to perhaps B_{50} (=495057205241079648212477525/66). It would have been reasonably fast too; the Analytical Engine was intended to do about 7 operations per second. So Ada's B_8 would have taken about 5 seconds and B_{50} would have taken perhaps a minute.

Curiously, even in our record-breaking computation of Bernoulli numbers a few years ago, we were basically using the same algorithm as Ada—though now there are slightly faster algorithms that effectively compute Bernoulli number numerators modulo a sequence of primes, then reconstruct the full numbers using the Chinese Remainder Theorem.

Babbage vs. Ada?

The Analytical Engine and its construction were all Babbage's work. So what did Ada add? Ada saw herself first and foremost as an expositor. Babbage had shown her lots of plans and examples of the Analytical Engine. She wanted to explain what the overall point was—as well as relate it, as she put it, to "large, general, & metaphysical views".

In the surviving archive of Babbage's papers (discovered years later in his lawyer's family's cowhide trunk), there are a remarkable number of drafts of expositions of the Analytical Engine, starting in the 1830s, and continuing for decades, with titles like "Of the Analytical Engine" and "The Science of Number Reduced to Mechanism". Why Babbage never published any of these isn't clear. They seem like perfectly decent descriptions of the basic operation of the engine—though they are definitely more pedestrian than what Ada produced.

When Babbage died, he was writing a "History of the Analytical Engine," which his son completed. In it, there is a dated list of "446 Notations of the Analytical Engine," each essentially a representation of how some operation—like division—could be done on the Analytical Engine. The dates start in the 1830s, and run through the mid-1840s, with not much happening in the summer of 1843.

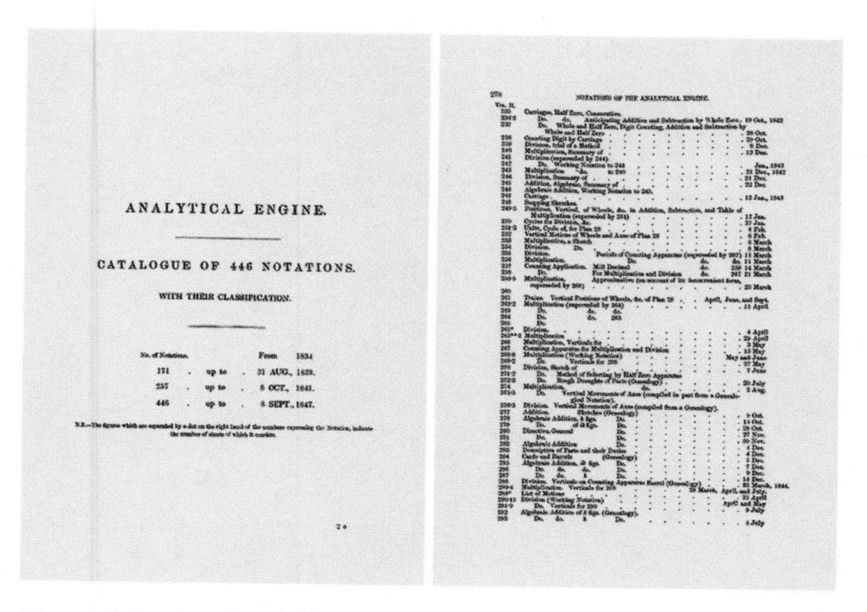

ANALYTICAL ENGINE.

CATALOGUE OF 446 NOTATIONS.

WITH THEIR CLASSIFICATION.

No. of Notations.		From 1834
171	up to	31 AUG., 1839.
257	up to	8 OCT., 1841.
446	up to	8 SEPT., 1847.

NOTATIONS OF THE ANALYTICAL ENGINE.

Meanwhile, in the collection of Babbage's papers at the Science Museum, there are some sketches of higher-level operations on the Analytical Engine. For example, from 1837 there's "Elimination between two equations of the first degree"—essentially the evaluation of a rational function:

There are a few very simple recurrence relations:

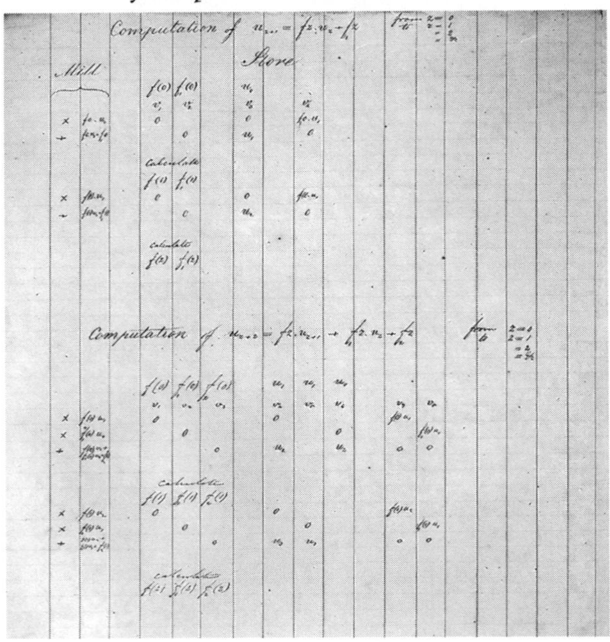

Then from 1838, there's a computation of the coefficients in the product of two polynomials:

Formula
Mill
Store
Results

But there's nothing as sophisticated—or as clean—as Ada's computation of the Bernoulli numbers. Babbage certainly helped and commented on Ada's work, but she was definitely the driver of it.

So what did Babbage say about that? In his autobiography written 26 years later, he had a hard time saying anything nice about anyone or anything. About Ada's Notes, he writes, "We discussed together the various illustrations that might be introduced: I suggested several, but the selection was entirely her own. So also was the algebraic working out of the different problems, except, indeed, that relating to the numbers of Bernoulli, which I had offered to do to save Lady Lovelace the trouble. This she sent back to me for an amendment, having detected a grave mistake which I had made in the process."

When I first read this, I thought Babbage was saying that he basically ghostwrote all of Ada's Notes. But reading what he wrote again, I realize it actually says almost nothing, other than that he suggested things that Ada may or may not have used.

To me, there's little doubt about what happened: Ada had an idea of what the Analytical Engine should be capable of, and was asking Babbage questions about how it could be achieved. If my own experiences with hardware designers in modern times are anything to go by, the answers will often have been very detailed. Ada's achievement was to distill from these details a clear exposition of the abstract operation of the machine—something which Babbage never did. (In his autobiography, he basically just refers to Ada's Notes.)

Babbage's Secret Sauce

For all his various shortcomings, the very fact that Babbage figured out how to build even a functioning Difference Engine—let alone an Analytical Engine—is extremely impressive. So how did he do it? I think the key was what he called his Mechanical Notation. He first wrote about it in 1826 under the title "On a Method of Expressing by Signs the Action of Machinery". His idea was to take a detailed structure of a machine and abstract a kind of symbolic diagram of how its parts act on each other. His first example was a hydraulic device:

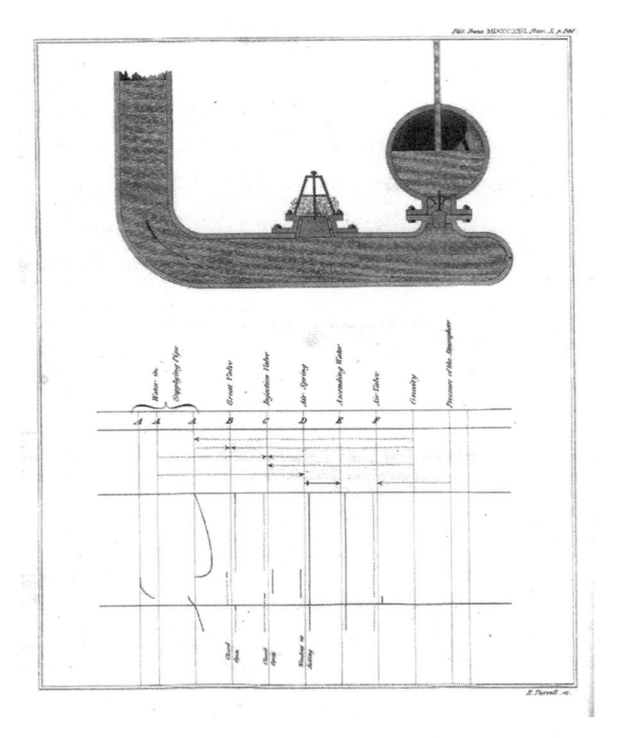

Then he gave the example of a clock, showing on the left a kind of "execution trace" of how the components of the clock change, and on the right a kind of "block diagram" of their relationships:

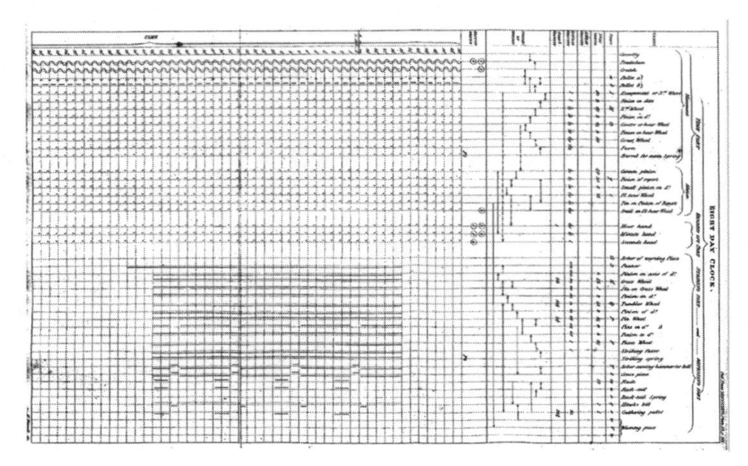

It's a pretty nice way to represent how a system works, similar in some ways to a modern timing diagram—but not quite the same. And over the years that Babbage worked on the Analytical Engine, his notes show ever more complex diagrams. It's not quite clear what something like this means:

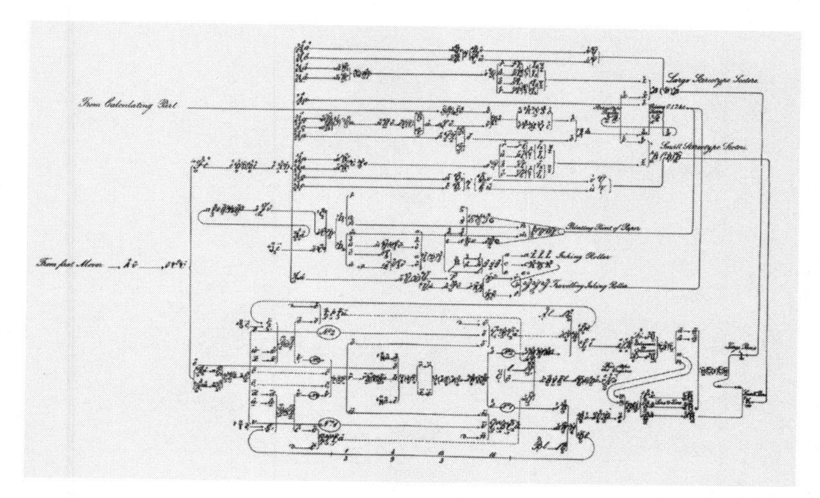

But it looks surprisingly like a modern Modelica representation—say in Wolfram SystemModeler. (One difference in modern times is that subsystems are represented much more hierarchically; another is that everything is now computable, so that actual behavior of the system can be simulated from the representation.)

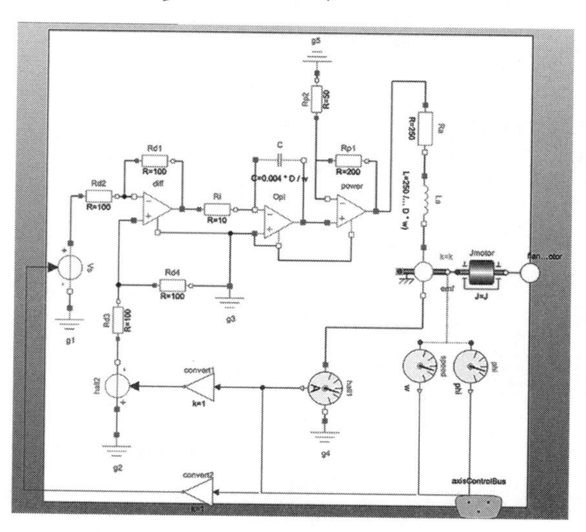

But even though Babbage used his various kinds of diagrams extensively himself, he didn't write papers about them. Indeed, his only other publication about "Mechanical Notation" is the flyer he had printed up for the Great Exhibition in 1851—apparently a pitch for standardization in drawings of mechanical components (and indeed these notations appear on Babbage's diagrams like the one above).

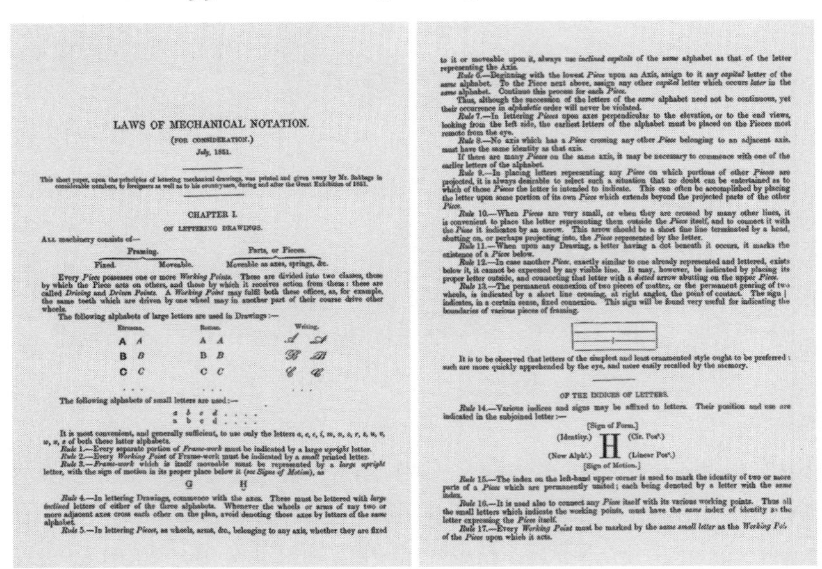

LAWS OF MECHANICAL NOTATION.

(FOR CONSIDERATION.)

July, 1851.

This short paper, upon the principles of lettering mechanical drawings, was printed and given away by Mr. Babbage in considerable numbers, to foreigners as well as to his countrymen, during and after the Great Exhibition of 1851.

CHAPTER I.

ON LETTERING DRAWINGS.

ALL machinery consists of—

Framing: Fixed. Moveable.
Parts, or Pieces: Moveable as axes, springs, &c.

Every *Piece* possesses one or more *Working Points*. These are divided into two classes, those by which the Piece acts on others, and those by which it receives action from them: these are called *Driving* and *Driven Points*. A *Working Point* may fulfil both these offices, as, for example, the same teeth which are driven by one wheel may in another part of their course drive other wheels.

The following alphabets of large letters are used in Drawings:—

Etruscan.	Roman.	Writing.
A *A*	A *A*	𝒜 𝒜
B *B*	B *B*	ℬ ℬ
C *C*	C *C*	𝒞 𝒞
. . .	. . .	. . .

The following alphabets of small letters are used:—

a b c d . . . .
a b c d . . . .

It is most convenient, and generally sufficient, to use only the letters *a, c, e, i, m, n, o, r, s, u, v, w, x, z* of both these latter alphabets.

Rule 1.—Every separate portion of *Frame-work* must be indicated by a large *upright* letter.

Rule 2.—Every *Working Point* of Frame-work must be indicated by a *small* printed letter.

Rule 3.—*Frame-work* which is itself moveable must be represented by a *large upright* letter, with the sign of motion in its proper place below it (*see Signs of Motion*), as

G H

Rule 4.—In lettering Drawings, commence with the axes. These must be lettered with *large inclined* letters of either of the three alphabets. Whenever the wheels or arms of any two or more adjacent axes cross each other on the plan, avoid denoting those axes by letters of the same alphabet.

Rule 5.—In lettering *Pieces*, as wheels, arms, &c., belonging to any axis, whether they are fixed to it or moveable upon it, always use *inclined capitals* of the *same* alphabet as that of the letter representing the Axis.

Rule 6.—Beginning with the lowest *Piece* upon an Axis, assign to it any *capital* letter of the *same* alphabet. To the Piece next above, assign any other *capital* letter which occurs *later* in the *same* alphabet. Continue this process for each *Piece*.

Thus, although the succession of the letters of the *same* alphabet need not be continuous, yet their occurrence in *alphabetic* order will never be violated.

Rule 7.—In lettering *Pieces* upon axes perpendicular to the elevation, or to the end views, looking from the left side, the earliest letters of the alphabet must be placed on the Pieces most remote from the eye.

Rule 8.—No axis which has a *Piece* crossing any other *Piece* belonging to an adjacent axis, must have the same identity as that axis.

If there are many *Pieces* on the same axis, it may be necessary to commence with one of the earlier letters of the alphabet.

Rule 9.—In placing letters representing any *Piece* on which portions of other *Pieces* are projected, it is always desirable to select such a situation that no doubt can be entertained as to which of those *Pieces* the letter is intended to indicate. This can often be accomplished by placing the letter upon some portion of its own *Piece* which extends beyond the projected parts of the other *Piece*.

Rule 10.—When *Pieces* are very small, or when they are crossed by many other lines, it is convenient to place the letter representing them outside the *Piece* itself, and to connect it with the *Piece* it indicates by an arrow. This arrow should be a short fine line terminated by a head, abutting on, or perhaps projecting into, the *Piece* represented by the letter.

Rule 11.—When upon any Drawing, a letter having a dot beneath it occurs, it marks the existence of a *Piece* below.

Rule 12.—In case another *Piece*, exactly similar to one already represented and lettered, exists below it, it cannot be expressed by any visible line. It may, however, be indicated by placing its proper letter outside, and connecting that letter with a *dotted* arrow abutting on the upper *Piece*.

Rule 13.—The permanent connexion of two pieces of matter, or the permanent gearing of two wheels, is indicated by a short line crossing, at right angles, the point of contact. The sign | indicates, in a certain sense, fixed connexion. This sign will be found very useful for indicating the boundaries of various pieces of framing.

It is to be observed that letters of the simplest and least ornamented style ought to be preferred: such are more quickly apprehended by the eye, and more easily recalled by the memory.

OF THE INDICES OF LETTERS.

Rule 14.—Various indices and signs may be affixed to letters. Their position and use are indicated in the subjoined letter:—

[Sign of Form.]
(Identity.) H (Cir. Posn.)
(New Alpht.) (Linear Posn.)
[Sign of Motion.]

Rule 15.—The index on the left-hand upper corner is used to mark the identity of two or more parts of a *Piece* which are permanently united; each being denoted by a letter with the *same* index.

Rule 16.—It is used also to connect any *Piece* itself with its various working points. Thus all the small letters which indicate the working points, must have the *same* index of identity as the letter expressing the *Piece* itself.

Rule 17.—Every *Working Point* must be marked by the *same small letter* as the *Working Poi[nt]* of the *Piece* upon which it acts.

I'm not sure why Babbage didn't do more to explain his Mechanical Notation and his diagrams. Perhaps he was just bitter about people's failure to appreciate it in 1826. Or perhaps he saw it as the secret that let him create his designs. And even though systems engineering has progressed a long way since Babbage's time, there may yet be inspiration to be had from what Babbage did.

The Bigger Picture

OK, so what's the bigger picture of what happened with Ada, Babbage and the Analytical Engine?

Charles Babbage was an energetic man who had many ideas, some of them good. At the age of 30 he thought of making mathematical tables by machine, and continued to pursue this idea until he died 49 years later, inventing the Analytical Engine as a way to achieve his objective.

He was good—even inspired—at the engineering details. He was bad at keeping a project on track.

Ada Lovelace was an intelligent woman who became friends with Babbage (there's zero evidence they were ever romantically involved). As something of a favor to Babbage, she wrote an exposition of the Analytical Engine, and in doing so she developed a more abstract understanding of it than Babbage had—and got a glimpse of the incredibly powerful idea of universal computation.

The Difference Engine and things like it are special-purpose computers, with hardware that's built to do only one kind of thing. One might have thought that to do lots of different kinds of things would necessarily require lots of different kinds of computers. But this isn't true. And instead it's a fundamental fact that it's possible to make general-purpose computers, where a single fixed piece of hardware can be programmed to do any computation. And it's this idea of universal computation that for example makes software possible—and that launched the whole computer revolution in the 20th century.

Gottfried Leibniz had already had a philosophical concept of something like universal computation back in the 1600s. But it wasn't followed up. And Babbage's Analytical Engine is the first explicit example we know of a machine that would have been capable of universal computation.

Babbage didn't think of it in these terms, though. He just wanted a machine that was as effective as possible at producing mathematical tables. But in the effort to design this, he ended up with a universal computer.

When Ada wrote about Babbage's machine, she wanted to explain what it did in the clearest way—and to do this she looked at the machine more abstractly, with the result that she ended up exploring and articulating something quite recognizable as the modern notion of universal computation.

What Ada did was lost for many years. But as the field of mathematical logic developed, the idea of universal computation arose again, most clearly in the work of Alan Turing in 1936. Then when electronic computers were built in the 1940s, it was realized they too exhibited universal computation, and the connection was made with Turing's work.

There was still, though, a suspicion that perhaps some other way of making computers might lead to a different form of computation. And it actually wasn't until the 1980s that universal computation became widely accepted as a robust notion. And by that time, something new was emerging—notably through work I was doing: that universal computation was not only something that's possible, but that it's actually common.

And what we now know (embodied for example in my Principle of Computational Equivalence) is that beyond a low threshold a very wide range of systems—even of very simple construction—are actually capable of universal computation.

A Difference Engine doesn't get there. But as soon as one adds just a little more, one will have universal computation. So in retrospect, it's not surprising that the Analytical Engine was capable of universal computation.

Today, with computers and software all around us, the notion of universal computation seems almost obvious: of course we can use software to compute anything we want. But in the abstract, things might not be that way. And I think one can fairly say that Ada Lovelace was the first person ever to glimpse with any clarity what has become a defining phenomenon of our technology and even our civilization: the notion of universal computation.

What If...?

What if Ada's health hadn't failed—and she had successfully taken over the Analytical Engine project? What might have happened then?

I don't doubt that the Analytical Engine would have been built. Maybe Babbage would have had to revise his plans a bit, but I'm sure he would have made it work. The thing would have been the size of a train locomotive, with maybe 50,000 moving parts. And no doubt it would have been able to compute mathematical tables to 30- or 50-digit precision at the rate of perhaps one result every 4 seconds.

Would they have figured that the machine could be electromechanical rather than purely mechanical? I suspect so. After all, Charles Wheatstone, who was intimately involved in the development of the electric telegraph in the 1830s, was a good friend of theirs. And by transmitting information electrically through wires, rather than mechanically through rods, the hardware for the machine would have been dramatically reduced, and its reliability (which would have been a big issue) would have been dramatically increased.

Another major way that modern computers reduce hardware is by dealing with numbers in binary rather than decimal. Would they have figured that idea out? Leibniz knew about binary. And if George Boole had followed up on his meeting with Babbage at the Great Exhibition, maybe that would have led to something. Binary wasn't well known in the mid-1800s, but it did appear in puzzles, and Babbage, at least, was quite into puzzles: a notable example being his question of how to make a square of words with "bishop" along the top and side (which now takes just a few lines of Wolfram Language code to solve).

Babbage's primary conception of the Analytical Engine was as a machine for automatically producing mathematical tables—either printing them out by typesetting, or giving them as plots by drawing onto a plate. He imagined that humans would be the main users of these tables—although he did think of the idea of having libraries of pre-computed cards that would provide machine-readable versions.

Today—in the Wolfram Language for example—we never store much in the way of mathematical tables; we just compute what we need when we need it. But in Babbage's day—with the idea of a massive Analytical Engine—this way of doing things would have been unthinkable.

So, OK: would the Analytical Engine have gotten beyond computing mathematical tables? I suspect so. If Ada had lived as long as Babbage, she would still have been around in the 1890s when Herman Hollerith was doing card-based electromechanical tabulation for the census (and founding what would eventually become IBM). The Analytical Engine could have done much more.

Perhaps Ada would have used the Analytical Engine—as she began to imagine—to produce algorithmic music. Perhaps they would have used it to solve things like the three-body problem, maybe even by simulation. If they'd figured out binary, maybe they would even have simulated things like cellular automata.

Neither Babbage nor Ada ever made money commercially (and, as Babbage took pains to point out, his government contracts just paid his engineers, not him). If they had developed the Analytical Engine, would they have found a business model for it? No doubt they would have sold some engines to governments. Maybe they would even have operated a kind of cloud computing service for Victorian science, technology, finance and more.

But none of this actually happened, and instead Ada died young, the Analytical Engine was never finished, and it took until the 20th century for the power of computation to be discovered.

What Were They Like?

If one had met Charles Babbage, what would he have been like? He was, I think, a good conversationalist. Early in life he was idealistic ("do my best to leave the world wiser than I found it"); later he was almost a Dickensian caricature of a bitter old man. He gave good parties, and put great value in connecting with the highest strata of intellectual society. But particularly in his later years, he spent most of his time alone in his large house, filled with books and papers and unfinished projects.

Babbage was never a terribly good judge of people, or of how what he said would be perceived by them. And even in his eighties, he was still quite child-like in his polemics. He was also notoriously poor at staying focused; he always had a new idea to pursue. The one big exception to this was his almost-50-year persistence in trying to automate the process of computation.

I myself have shared a modern version of this very goal in my own life (..., Mathematica, Wolfram|Alpha, Wolfram Language, ...)—though so far only for 40 years. I am fortunate to have lived in a time when ambient technology made this much easier to achieve, but in every large project I have done it has still taken a certain singlemindedness and gritty tenacity—as well as leadership—to actually get it finished.

So what about Ada? From everything I can tell, she was a clear speaking, clear thinking individual. She came from the upper classes, but didn't wear especially fashionable clothes, and carried herself much less like a stereotypical countess than like an intellectual. As an adult, she was emotionally quite mature—probably more so than Babbage—and seems to have had a good practical grasp of people and the world.

Like Babbage, she was independently wealthy, and had no need to work for a living. But she was ambitious, and wanted to make something of herself. In person, beyond the polished Victorian upper-class exterior, I suspect she was something of a nerd, complete with math jokes and everything. She was also capable of great and sustained focus, for example over the months she spent writing her Notes.

In mathematics, she successfully learned up to the state of the art in her time—probably about the same level as Babbage. Unlike Babbage, we don't know of any specific research she did in mathematics, so it's hard to judge how good she would have been; Babbage was respectable though unremarkable.

When one reads Ada's letters, what comes through is a smart, sophisticated person, with a clear, logical mind. What she says is often dressed in Victorian pleasantries—but underneath, the ideas are clear and often quite forceful.

Ada was very conscious of her family background, and of being "Lord Byron's daughter". At some level, his story and success no doubt fueled her ambition, and her willingness to try new things. (I can't help thinking of her leading the engineers of the Analytical Engine as a bit like Lord Byron leading the Greek army.) But I also suspect his troubles loomed over her. For many years, partly at her mother's behest, she eschewed things like poetry. But she was drawn to abstract ways of thinking, not only in mathematics and science, but also in more metaphysical areas.

And she seems to have concluded that her greatest strength would be in bridging the scientific with the metaphysical—perhaps in what she called "poetical science". It was likely a correct self perception. For that is in a sense exactly what she did in the Notes she wrote: she took Babbage's detailed engineering, and made it more abstract and "metaphysical"—and in the process gave us a first glimpse of the idea of universal computation.

The Final Story

The story of Ada and Babbage has many interesting themes. It is a story of technical prowess meeting abstract "big picture" thinking. It is a story of friendship between old and young. It is a story of people who had the confidence to be original and creative.

It is also a tragedy. A tragedy for Babbage, who lost so many people in his life, and whose personality pushed others away and prevented him from realizing his ambitions. A tragedy for Ada, who was just getting started in something she loved when her health failed.

We will never know what Ada could have become. Another Mary Somerville, famous Victorian expositor of science? A Steve-Jobs-like figure who would lead the vision of the Analytical Engine? Or an Alan Turing, understanding the abstract idea of universal computation?

That Ada touched what would become a defining intellectual idea of our time was good fortune. Babbage did not know what he had; Ada started to see glimpses and successfully described them.

For someone like me the story of Ada and Babbage has particular resonance. Like Babbage, I have spent much of my life pursuing particular goals—though unlike Babbage, I have been able to see a fair fraction of them achieved. And I suspect that, like Ada, I have been put in a position where I can potentially see glimpses of some of the great ideas of the future.

But the challenge is to be enough of an Ada to grasp what's there—or at least to find an Ada who does. But at least now I think I have an idea of what the original Ada born 200 years ago today was like: a fitting personality on the road to universal computation and the present and future achievements of computational thinking.

It's been a pleasure getting to know you, Ada.

Quite a few organizations and people helped in getting information and material for this piece. I'd like to thank the British Library; the Museum of the History of Science, Oxford, Science Museum, London; the Bodleian Library, Oxford (with permission from the Earl of Lytton, Ada's great-great grandson, and one of her 10 living descendants); the New York Public Library; St. Mary Magdalene Church, Hucknall, Nottinghamshire (Ada's burial place); and Betty Toole (author of a collection of Ada's letters); as well as two old friends: Tim Robinson (re-creator of Babbage engines) and Nathan Myhrvold (funder of Difference Engine #2 re-creation).

Gottfried Leibniz

May 14, 2013

I've been curious about Gottfried Leibniz for years, not least because he seems to have wanted to build something like Mathematica and Wolfram|Alpha, and perhaps *A New Kind of Science* as well—though three centuries too early. So when I took a trip recently to Germany, I was excited to be able to visit his archive in Hanover.

Leafing through his yellowed (but still robust enough for me to touch) pages of notes, I felt a certain connection—as I tried to imagine what he was thinking when he wrote them, and tried to relate what I saw in them to what we now know after three more centuries:

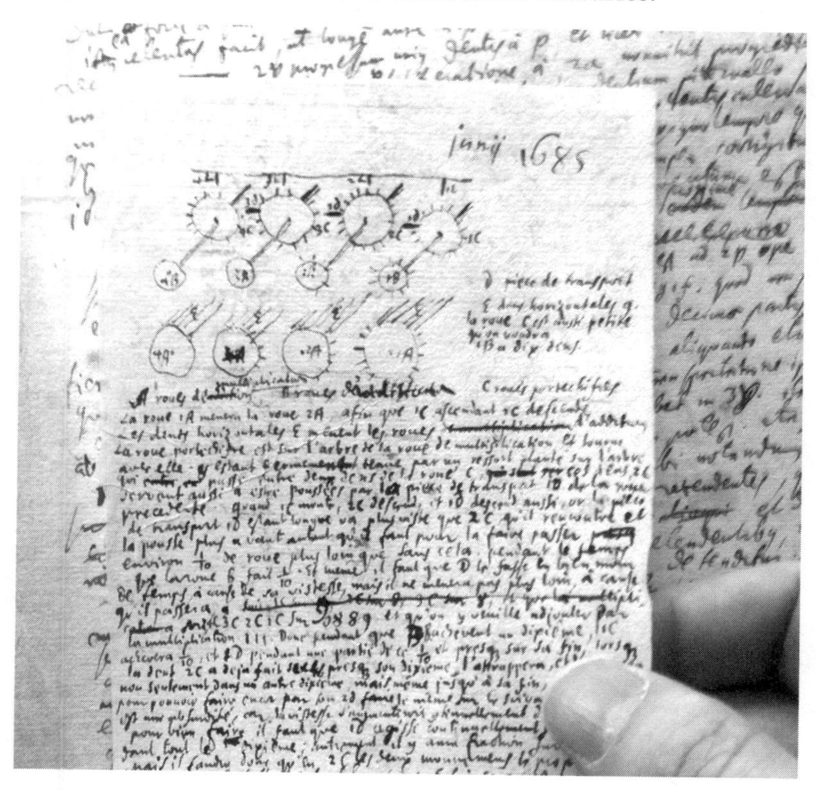

Some things, especially in mathematics, are quite timeless. Like here's Leibniz writing down an infinite series for $\sqrt{2}$ (the text is in Latin):

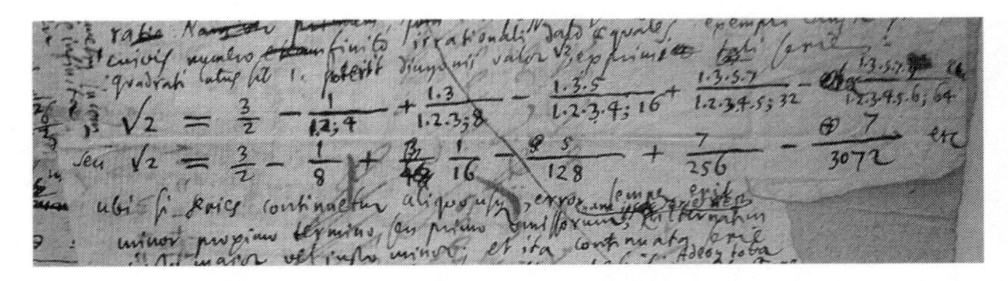

Or here's Leibniz trying to calculate a continued fraction—though he got the arithmetic wrong, even though he wrote it all out (the Π was his earlier version of an equal sign):

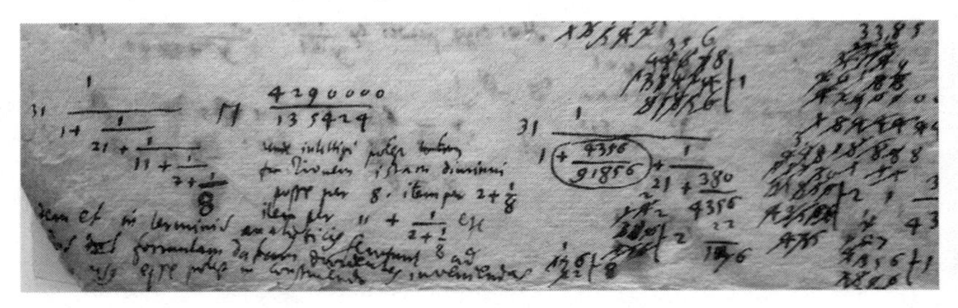

Or here's a little summary of calculus, that could almost be in a modern textbook:

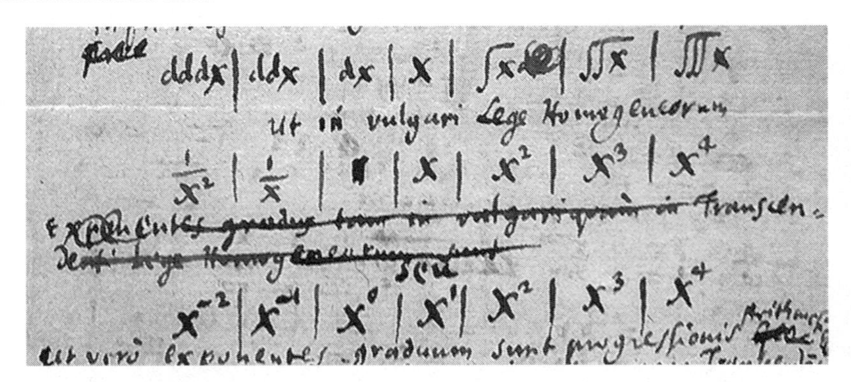

But what was everything else about? What was the larger story of his work and thinking?

I have always found Leibniz a somewhat confusing figure. He did many seemingly disparate and unrelated things—in philosophy, mathematics, theology, law, physics, history, and more. And he described what he was doing in what seem to us now as strange 17th-century terms.

But as I've learned more, and gotten a better feeling for Leibniz as a person, I've realized that underneath much of what he did was a core intellectual direction that is curiously close to the modern computational one that I, for example, have followed.

Gottfried Leibniz was born in Leipzig in what's now Germany in 1646 (four years after Galileo died, and four years after Newton was born). His father was a professor of philosophy; his mother's family was in the book trade. Leibniz's father died when Leibniz was 6—and after a 2-year deliberation on its suitability for one so young, Leibniz was allowed into his father's library, and began to read his way through its diverse collection of books. He went to the local university at age 15, studying philosophy and law—and graduated in both of them at age 20.

Even as a teenager, Leibniz seems to have been interested in systematization and formalization of knowledge. There had been vague ideas for a long time—for example in the semi-mystical *Ars Magna* of Ramon Llull from the 1300s—that one might be able to set up some kind of universal system in which all knowledge could be derived from combinations of signs drawn from a suitable (as Descartes called it) "alphabet of human thought". And for his philosophy graduation thesis, Leibniz tried to pursue this idea. He used some basic combinatorial mathematics to count possibilities. He talked about decomposing ideas into simple components on which a "logic of invention" could operate. And, for good measure, he put in an argument that purported to prove the existence of God.

As Leibniz himself said in later years, this thesis—written at age 20—was in many ways naive. But I think it began to define Leibniz's lifelong way of thinking about all sorts of things. And so, for example, Leibniz's law graduation thesis about "perplexing legal cases" was all about how such cases could potentially be resolved by reducing them to logic and combinatorics.

Leibniz was on a track to become a professor, but instead he decided to embark on a life working as an advisor for various courts and political rulers. Some of what he did for them was scholarship, tracking down abstruse—but politically important—genealogy and history. Some of it was organization and systematization—of legal codes, libraries and so on. Some of it was practical engineering—like trying to work out better ways to keep water out of silver mines. And some of it—particularly in earlier years—was "on the ground" intellectual support for political maneuvering.

One such activity in 1672 took Leibniz to Paris for four years—during which time he interacted with many leading intellectual lights. Before then, Leibniz's knowledge of mathematics had been fairly basic. But in Paris he had the opportunity to learn all the latest ideas and methods. And for example he sought out Christiaan Huygens, who agreed to teach Leibniz mathematics—after he succeeded in passing the test of finding the sum of the reciprocals of the triangular numbers.

Over the years, Leibniz refined his ideas about the systematization and formalization of knowledge, imagining a whole architecture for how knowledge would—in modern terms—be made computational. He saw the first step as being the development of an *ars characteristica*—a methodology for assigning signs or symbolic representations to things, and in effect creating a uniform "alphabet of thought". And he then imagined—in remarkable resonance with what we now know about computation—that from this uniform representation it would be possible to find "truths of reason in any field... through a calculus, as in arithmetic or algebra".

He talked about his ideas under a variety of rather ambitious names like *scientia generalis* ("general method of knowledge"), *lingua philosophica* ("philosophical language"), *mathematique universelle* ("universal mathematics"), *characteristica universalis* ("universal system") and *calculus ratiocinator* ("calculus of thought"). He imagined applications ultimately in all areas—science, law, medicine, engineering, theology and more. But the one area in which he had clear success quite quickly was mathematics.

To me it's remarkable how rarely in the history of mathematics that notation has been viewed as a central issue. It happened at the beginning of modern mathematical logic in the late 1800s with the work of people like Gottlob Frege and Giuseppe Peano. And in recent times it's happened with me in my efforts to create Mathematica and the Wolfram Language. But it also happened three centuries ago with Leibniz. And I suspect that Leibniz's successes in mathematics were in no small part due to the effort he put into notation, and the clarity of reasoning about mathematical structures and processes that it brought.

When one looks at Leibniz's papers, it's interesting to see his notation and its development. Many things look quite modern. Though there are charming dashes of the 17th century, like the occasional use of alchemical or planetary symbols for algebraic variables:

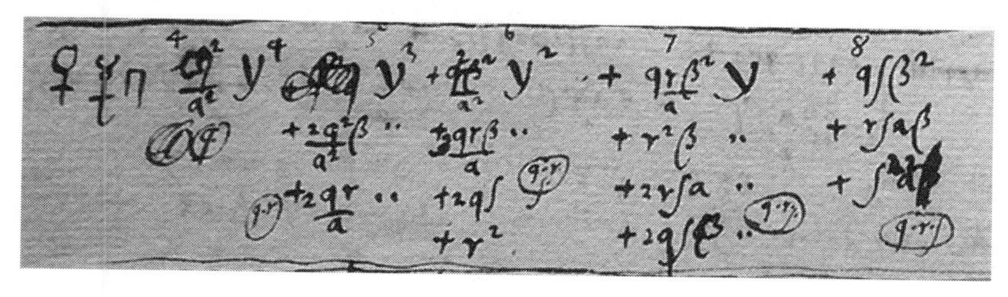

There's ∏ as an equals sign instead of =, with the slightly hacky idea of having it be like a balance, with a longer leg on one side or the other indicating less than (“<”) or greater than (“>”):

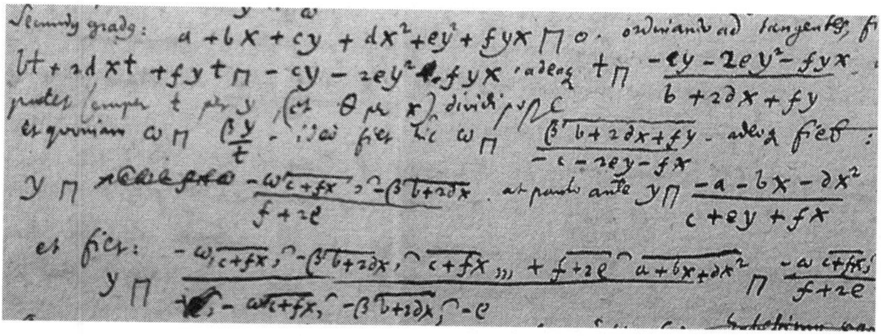

There are overbars to indicate grouping of terms—arguably a better idea than parentheses, though harder to type, and typeset:

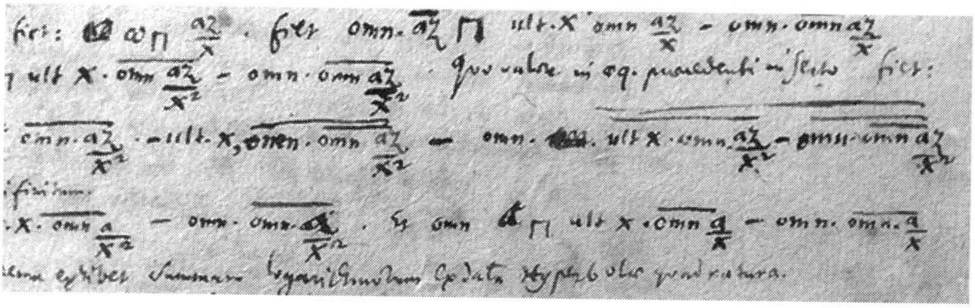

We do use overbars for roots today. But Leibniz wanted to use them in integrals too—along with the rather nice "tailed d," which reminds me of the double-struck "differential d" that we invented for representing integrals in Mathematica.

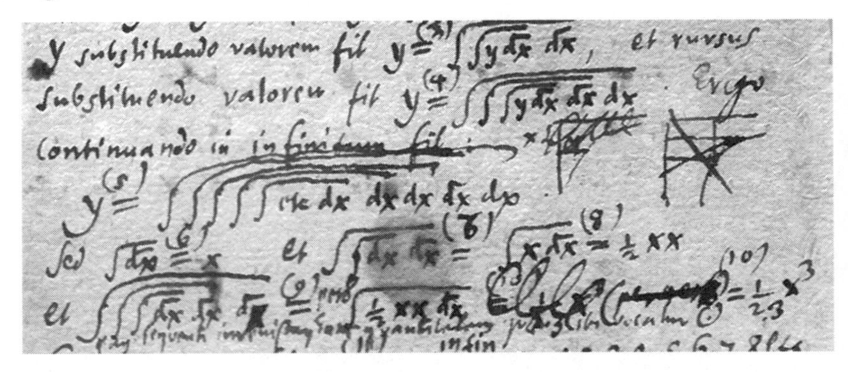

Particularly in solving equations, it's quite common to want to use ±, and it's always confusing how the grouping is supposed to work, say in $a \pm b \pm c$. Well, Leibniz seems to have found it confusing too, but he invented a notation to handle it—which we actually should consider using today too:

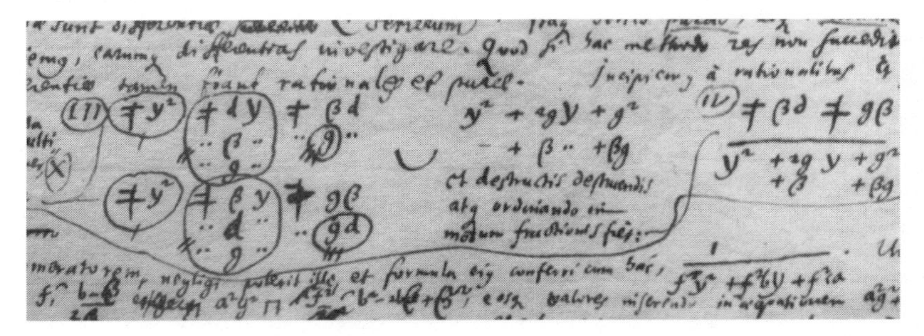

I'm not sure what some of Leibniz's notation means—though those overtildes are rather nice-looking:

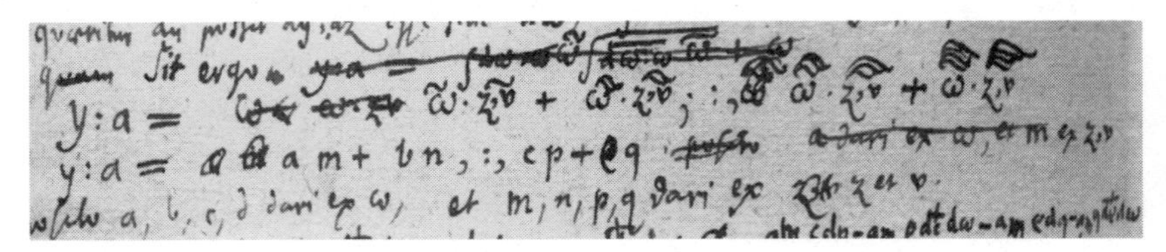

As are these things with dots:

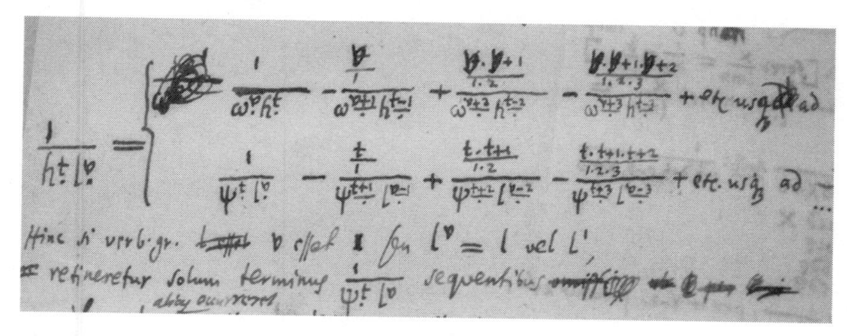

Or this interesting-looking diagrammatic form:

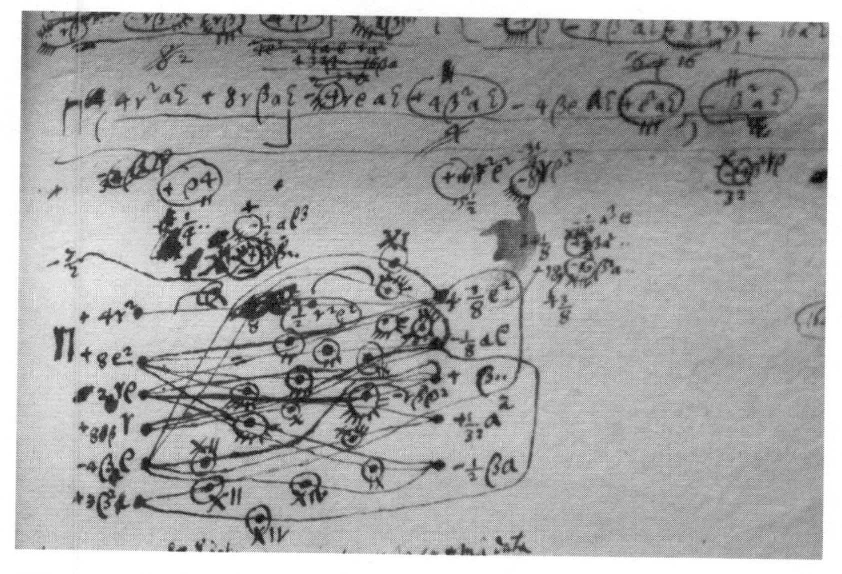

Of course, Leibniz's most famous notations are his integral sign (long "s" for "summa") and d, here summarized in the margin for the first time, on November 11, 1675 (the "5" in "1675" was changed to a "3" after the fact, perhaps by Leibniz):

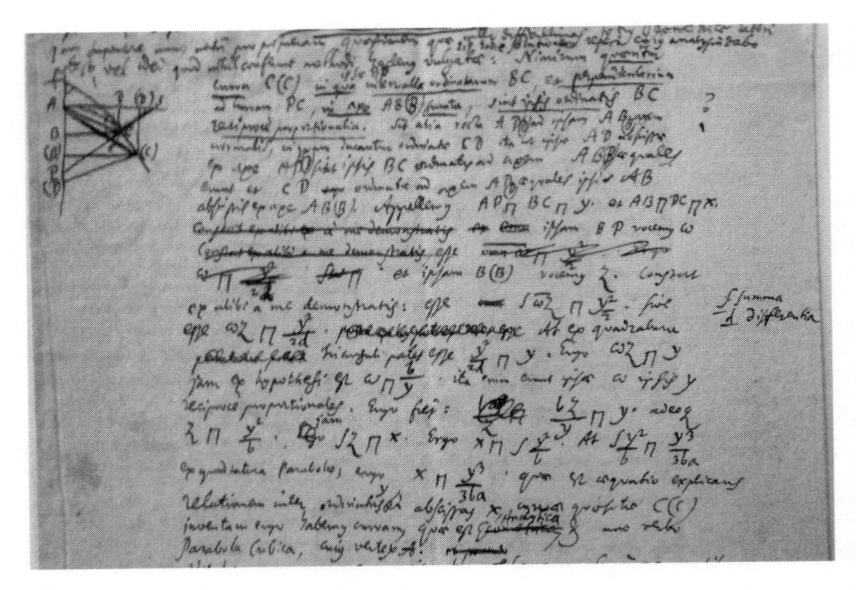

I find it interesting that despite all his notation for "calculational" operations, Leibniz apparently did not invent similar notation for logical operations. "Or" was just the Latin word *vel*, "and" was *et*, and so on. And when he came up with the idea of quantifiers (modern ∀ and ∃), he just represented them by the Latin abbreviations U.A. and P.A.:

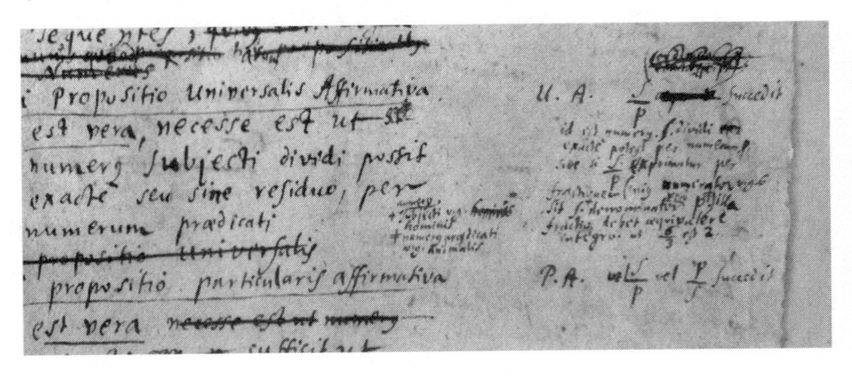
Propositio Universalis Affirmativa
est vera, necesse est ut
numerus subjecti dividi possit
exacte seu sine residuo, per
numerum praedicati
propositio particularis affirmativa
est vera
U. A.
P. A.

It's always struck me as a remarkable anomaly in the history of thought that it took until the 1930s for the idea of universal computation to emerge. And I've often wondered if lurking in the writings of Leibniz there might be an early version of universal computation—

maybe even a diagram that we could now interpret as a system like a Turing machine. But with more exposure to Leibniz, it's become clearer to me why that's probably not the case.

One big piece, I suspect, is that he didn't take discrete systems quite seriously enough. He referred to results in combinatorics as "self-evident", presumably because he considered them directly verifiable by methods like arithmetic. And it was only "geometrical", or continuous, mathematics that he felt needed to have a calculus developed for it. In describing things like properties of curves, Leibniz came up with something like continuous functions. But he never seems to have applied the idea of functions to discrete mathematics—which might for example have led him to think about universal elements for building up functions.

Leibniz recognized the success of his infinitesimal calculus, and was keen to come up with similar "calculi" for other things. And in another "near miss" with universal computation, Leibniz had the idea of encoding logical properties using numbers. He thought about associating every possible attribute of a thing with a different prime number, then characterizing the thing by the product of the primes for its attributes—and then representing logical inference by arithmetic operations. But he only considered static attributes—and never got to an idea like Gödel numbering where operations are also encoded in numbers.

But even though Leibniz did not get to the idea of universal computation, he did understand the notion that computation is in a sense mechanical. And indeed quite early in life he seems to have resolved to build an actual mechanical calculator for doing arithmetic. Perhaps in part it was because he wanted to use it himself (always a good reason to build a piece of technology!). For despite his prowess at algebra and the like, his papers are charmingly full of basic (and sometimes incorrect) school-level arithmetic calculations written out in the margin—and now preserved for posterity:

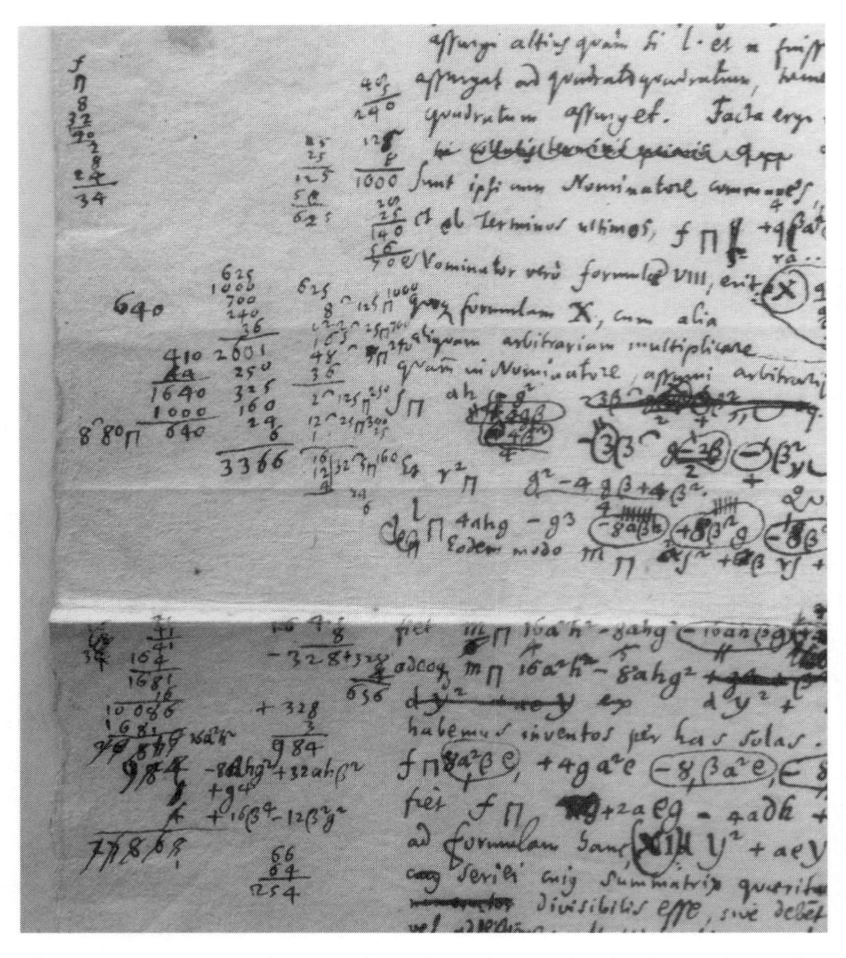

There were scattered examples of mechanical calculators being built in Leibniz's time, and when he was in Paris, Leibniz no doubt saw the addition calculator that had been built by Blaise Pascal in 1642. But Leibniz resolved to make a "universal" calculator, that could for the first time do all four basic functions of arithmetic with a single machine. And he wanted to give it a simple "user interface", where one would for example turn a handle one way for multiplication, and the opposite way for division.

In Leibniz's papers there are all sorts of diagrams about how the machine should work:

Leibniz imagined that his calculator would be of great practical utility—and indeed he seems to have hoped that he would be able to turn it into a successful business. But in practice, Leibniz struggled to get the calculator to work at all reliably. For like other mechanical calculators of its time, it was basically a glorified odometer. And just like in Charles Babbage's machines nearly 200 years later, it was mechanically difficult to make many wheels move at once when a cascade of carries occurred.

Leibniz at first had a wooden prototype of his machine built, intended to handle just 3 or 4 digits. But when he demoed this to people like Robert Hooke during a visit to London in 1673 it didn't go very well. But he kept on thinking he'd figured everything out—for example in 1679 writing (in French) of the "last correction to the arithmetic machine":

24 Aoust. 1679.

Dernière correction de la Machine Arithmetiq

Notes from 1682 suggest that there were more problems, however:

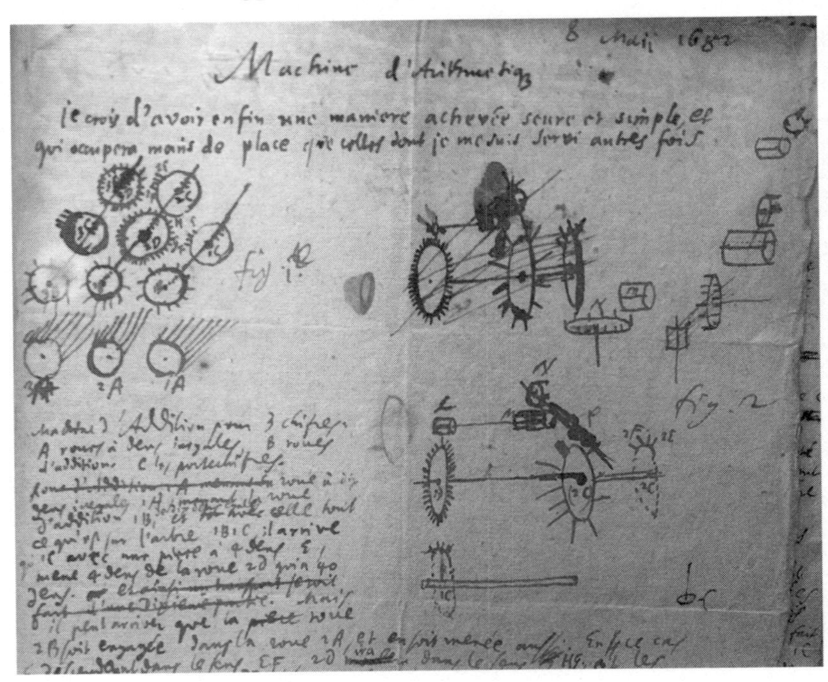

8 Maii 1682

Machine d'Arithmetiq

Je crois d'avoir enfin une maniere achevée seure et simple, et qui occupera moins de place que celles dont je me suis servi autres fois.

But Leibniz had plans drafted up from his notes—and contracted an engineer to build a brass version with more digits:

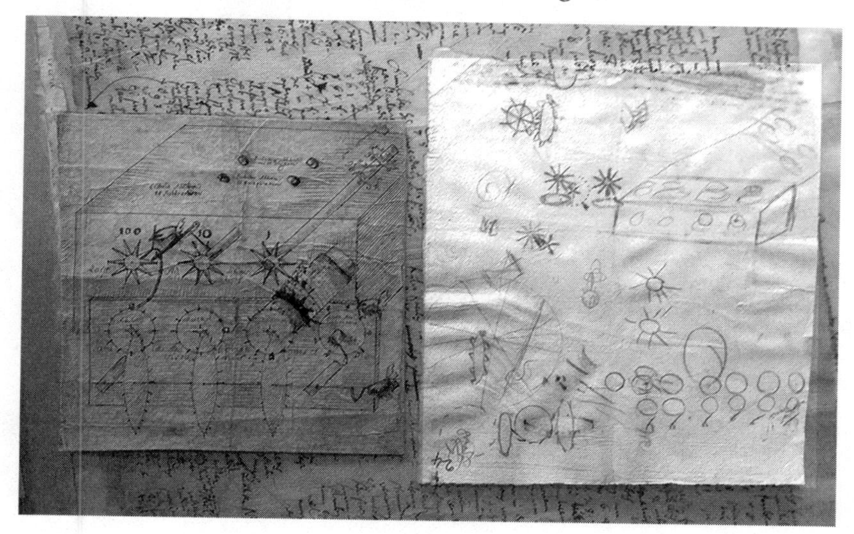

It's fun to see Leibniz's "marketing material" for the machine:

Machina Arithmetica.
in qua non additio tantum (et subtractio) sed
et multiplicatio nullo, divisio vero paene nullo,
animi labore peragantur.

Cum aliquot abhinc annis Instrumentum quoddam,
quod qui portat, passus ipse suos ne cogitans quidem
numerat; primum vidissem, statim subiit animum
cogitatio posse toti Arithmeticae simili machina-
menti genere subveniri, ut non numeratio tantum
sed et additio cum subtractione et multiplicatio
cum divisione, homine successus securo, ab ipsa
machina recte disposita, facile promteque

As well as parts of the “manual” (with 365×24 as a “worked example”):

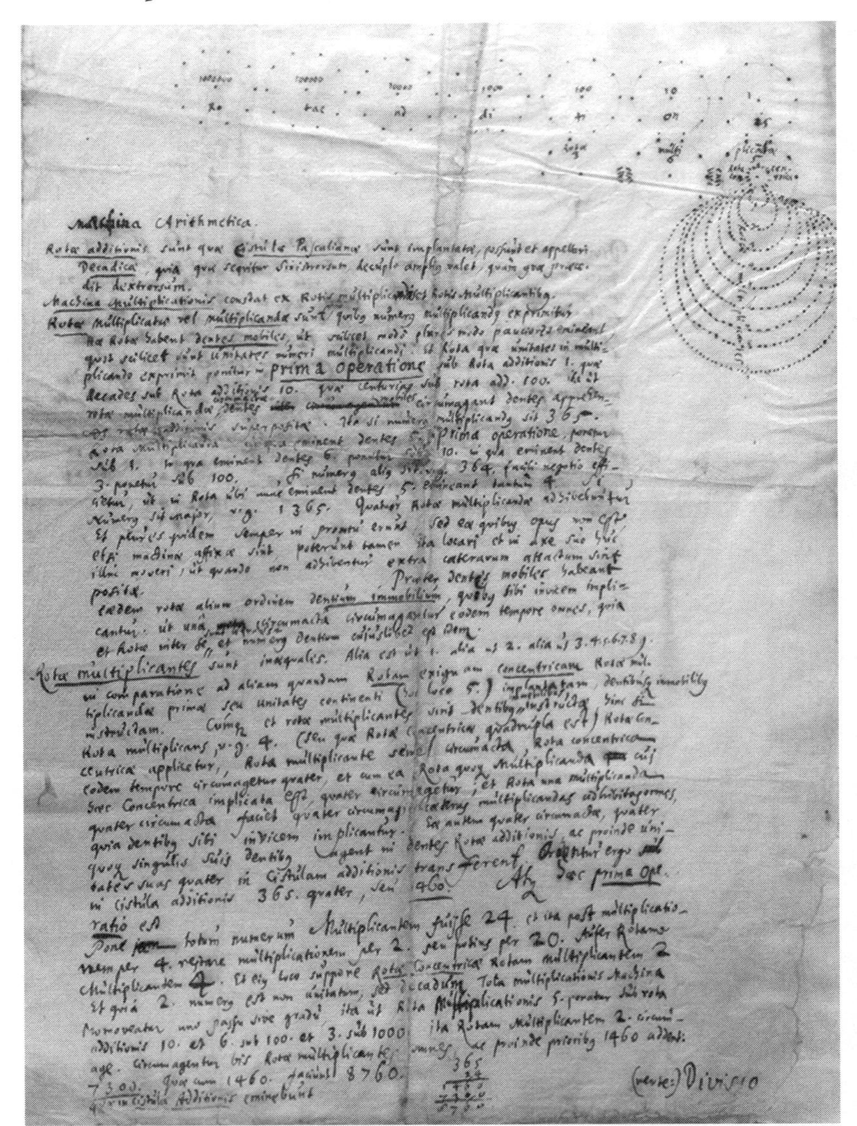

Machina Arithmetica.

Rotae additionis sunt quae similes Pascalianae sunt implantatae, possunt et appellari Decadicae, quia quae sequitur sinistrorsum decuplo amplius valet, quam quae praecedit dextrorsum.

Machina multiplicationis constat ex Rotis multiplicantibus et Rotis multiplicandis.

Complete with detailed usage diagrams:

But despite all this effort, problems with the calculator continued. And in fact, for more than 40 years, Leibniz kept on tweaking his calculator—probably altogether spending (in today's currency) more than a million dollars on it.

So what actually happened to the physical calculator? When I visited Leibniz's archive, I had to ask. "Well," my hosts said, "we can show you." And there in a vault, along with shelves of boxes, was Leibniz's calculator, looking as good as new in a glass case—here captured by me in a strange juxtaposition of ancient and modern:

All the pieces are there. Including a convenient wooden carrying box. Complete with a cranking handle. And, if it worked right, the ability to do any basic arithmetic operation with a few minutes of cranking:

Leibniz clearly viewed his calculator as a practical project. But he still wanted to generalize from it, for example trying to make a general "logic" to describe geometries of mechanical linkages. And he also thought about the nature of numbers and arithmetic. And was particularly struck by binary numbers.

Bases other than 10 had been used in recreational mathematics for several centuries. But Leibniz latched on to base 2 as having particular significance—and perhaps being a key bridge between philosophy, theology and mathematics. And he was encouraged in this by his realization that binary numbers were at the core of the *I Ching*, which he'd heard about from missionaries to China, and viewed as related in spirit to his *characteristica universalis*.

Leibniz worked out that it would be possible to build a calculator based on binary. But he appears to have thought that only base 10 could actually be useful.

It's strange to read what Leibniz wrote about binary numbers. Some of it is clear and practical—and still seems perfectly modern. But some of it is very 17th century—talking for example about how binary proves that everything can be made from nothing, with 1 being identified with God, and 0 with nothing.

Almost nothing was done with binary for a couple of centuries after Leibniz: in fact, until the rise of digital computing in the last few decades. So when one looks at Leibniz's papers, his calculations in binary are probably what seem most "out of his time":

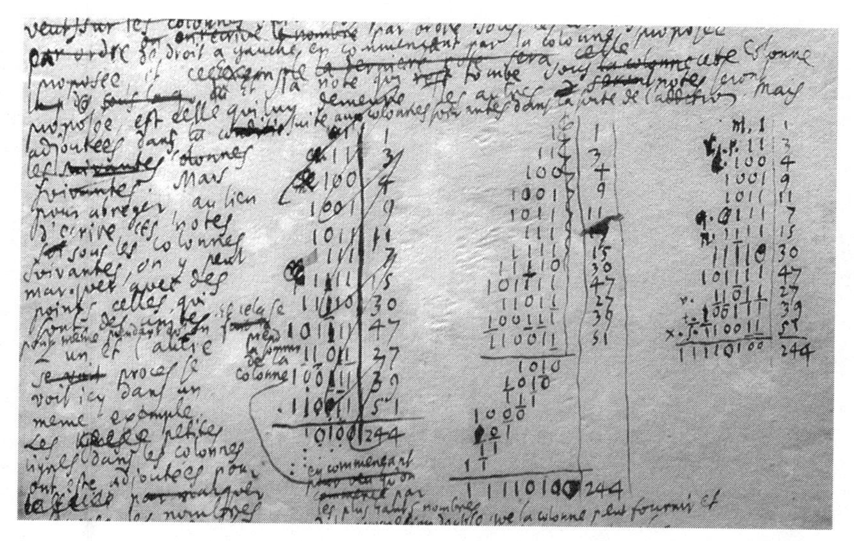

With binary, Leibniz was in a sense seeking the simplest possible underlying structure. And no doubt he was doing something similar when he talked about what he called "monads". I have to say that I've never really understood monads. And usually when I think I almost have, there's some mention of souls that just throws me completely off.

Still, I've always found it tantalizing that Leibniz seemed to conclude that the "best of all possible worlds" is the one "having the greatest variety of phenomena from the smallest number of principles". And indeed, in the prehistory of my work on *A New Kind of Science*, when I first started formulating and studying one-dimensional cellular automata in 1981, I considered naming them "polymones"—but at the last minute got cold feet when I got confused again about monads.

There's always been a certain mystique around Leibniz and his papers. Kurt Gödel—perhaps displaying his paranoia—seemed convinced that Leibniz had discovered great truths that had been suppressed for centuries. But while it is true that Leibniz's papers were sealed when he died, it was his work on topics like history and genealogy—and the state secrets they might entail—that was the concern.

Leibniz's papers were unsealed long ago, and after three centuries one might assume that every aspect of them would have been well studied. But the fact is that even after all this time, nobody has actually gone through all of the papers in full detail. It's not that there are so many of them. Altogether there are only about 200,000 pages—filling perhaps a dozen shelving units (and only a little larger than my own personal archive from just the 1980s). But the problem is the diversity of material. Not only lots of subjects. But also lots of overlapping drafts, notes and letters, with unclear relationships between them.

Leibniz's archive contains a bewildering array of documents. From the very large:

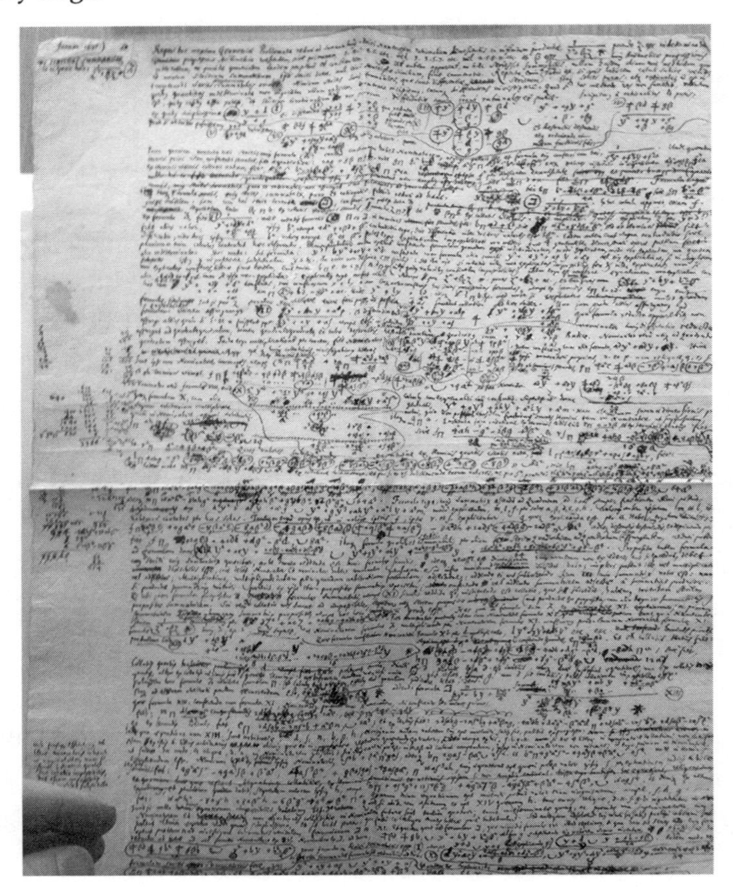

To the very small (Leibniz's writing got smaller as he got older and more near-sighted):

Most of the documents in the archive seem very serious and studious. But despite the high cost of paper in Leibniz's time, one still finds preserved for posterity the occasional doodle (is that Spinoza, by any chance?):

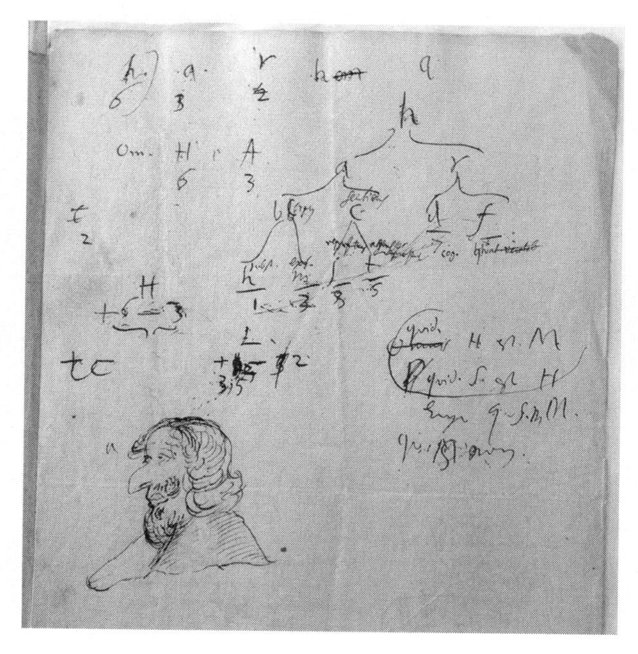

Leibniz exchanged mail with hundreds of people—famous and not-so-famous—all over Europe. So now, 300 years later, one can find in his archive "random letters" from the likes of Jacob Bernoulli:

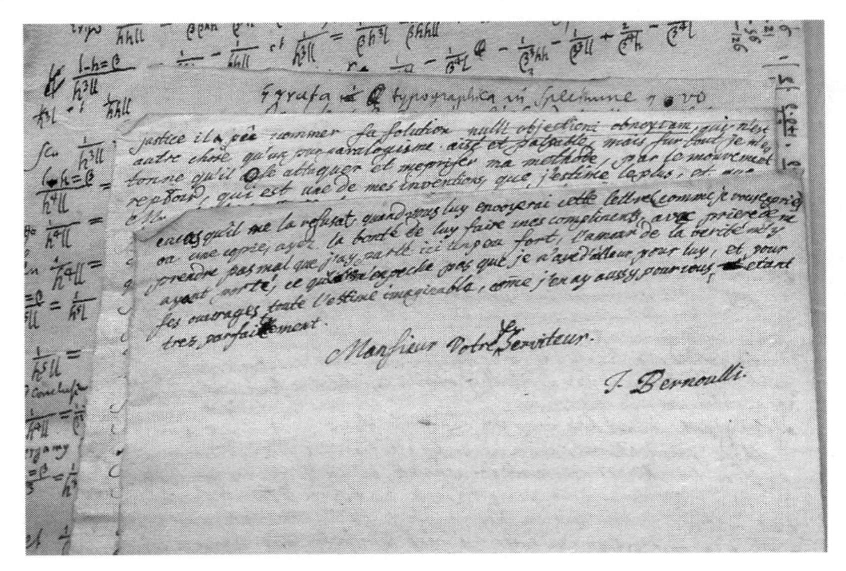

Monsieur Votre Serviteur.

J. Bernoulli.

What did Leibniz look like? Here he is, both in an official portrait, and without his rather oversized wig (that was mocked even in his time), that he presumably wore to cover up a large cyst on his head:

As a person, Leibniz seems to have been polite, courtierly and even-tempered. In some ways, he may have come across as something of a

nerd, expounding at great depth on all manner of topics. He seems to have taken great pains—as he did in his letters—to adapt to whoever he was talking to, emphasizing theology when he was talking to a theologian, and so on. Like quite a few intellectuals of his time, Leibniz never married, though he seems to have been something of a favorite with women at court.

In his career as a courtier, Leibniz was keen to climb the ladder. But not being into hunting or drinking, he never quite fit in with the inner circles of the rulers he worked for. Late in his life, when George I of Hanover became king of England, it would have been natural for Leibniz to join his court. But Leibniz was told that before he could go, he had to start writing up a history project he'd supposedly been working on for 30 years. Had he done so before he died, he might well have gone to England and had a very different kind of interaction with Newton.

At Leibniz's archive, there are lots of papers, his mechanical calculator, and one more thing: a folding chair that he took with him when he traveled, and that he had suspended in carriages so he could continue to write as the carriage moved:

Leibniz was quite concerned about status (he often styled himself "Gottfried von Leibniz", though nobody quite knew where the "von" came from). And as a form of recognition for his discoveries, he wanted to have a medal created to commemorate binary numbers. He came up with a detailed design, complete with the tag line *omnibus ex nihilo ducendis; sufficit unum* ("everything can be derived from nothing; all that is needed is 1"). But nobody ever made the medal for him.

In 2007, though, I wanted to come up with a 60th birthday gift for my friend Greg Chaitin, who has been a long-time Leibniz enthusiast. And so I thought: why not actually make Leibniz's medal? So we did. Though on the back, instead of the picture of a duke that Leibniz proposed, we put a Latin inscription about Greg's work.

And when I visited the Leibniz archive, I made sure to bring a copy of the medal, so I could finally put a real medal next to Leibniz's design:

It would have been interesting to know what pithy statement Leibniz might have had on his grave. But as it was, when Leibniz died at the age of 70, his political fates were at a low ebb, and no elaborate memorial was constructed. Still, when I was in Hanover, I was keen to see his grave—which turns out to carry just the simple Latin inscription "bones of Leibniz":

Across town, however, there's another commemoration of a sort—an outlet store for cookies that carry the name "Leibniz" in his honor:

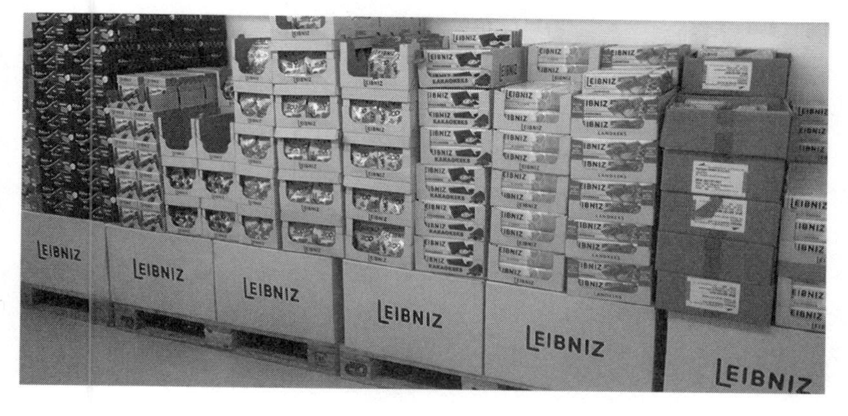

So what should we make of Leibniz in the end? Had history developed differently, there would probably be a direct line from Leibniz to modern computation. But as it is, much of what Leibniz tried to do stands isolated—to be understood mostly by projecting backward from modern computational thinking to the 17th century.

And with what we know now, it is fairly clear what Leibniz understood, and what he did not. He grasped the concept of having formal, symbolic representations for a wide range of different kinds of things. And he suspected that there might be universal elements (maybe even just 0 and 1) from which these representations could be built. And he understood that from a formal symbolic representation of knowledge, it should be possible to compute its consequences in mechanical ways—and perhaps create new knowledge by an enumeration of possibilities.

Some of what Leibniz wrote was abstract and philosophical—sometimes maddeningly so. But at some level Leibniz was also quite practical. And he had sufficient technical prowess to often be able to make real progress. His typical approach seems to have been to start by trying to create a formal structure to clarify things—with formal notation if possible. And after that his goal was to create some kind of "calculus" from which conclusions could systematically be drawn.

Realistically he only had true success with this in one specific area: continuous "geometrical" mathematics. It's a pity he never tried more seriously in discrete mathematics, because I think he might have been able to make progress, and might conceivably even have reached the idea of universal computation. He might well also have ended up starting to enumerate possible systems in the kind of way I have done in the computational universe.

One area where he did try his approach was with law. But in this he was surely far too early, and it is only now—300 years later—that computational law is beginning to seem realistic.

Leibniz also tried thinking about physics. But while he made progress with some specific concepts (like kinetic energy), he never managed to come up with any sort of large-scale "system of the world", of the kind that Newton in effect did in his *Principia*.

In some ways, I think Leibniz failed to make more progress because he was trying too hard to be practical, and—like Newton—to decode the operation of actual physics, rather than just looking at related formal structures. For had Leibniz tried to do at least the basic kinds of explorations that I did in *A New Kind of Science*, I don't think he would have had any technical difficulty—but I think the history of science could have been very different.

And I have come to realize that when Newton won the PR war against Leibniz over the invention of calculus, it was not just credit that was at stake, it was a way of thinking about science. Newton was in a sense quintessentially practical: he invented tools, then showed how these could be used to compute practical results about the physical world. But Leibniz had a broader and more philosophical view, and saw calculus not just as a specific tool in itself, but as an example that should inspire efforts at other kinds of formalization and other kinds of universal tools.

I have often thought that the modern computational way of thinking that I follow is somehow obvious—and somehow an inevitable feature of thinking about things in formal, structured ways. But it has never been very clear to me whether this apparent obviousness is just the result of modern times, and of our experience with modern practical computer technology. But looking at Leibniz, we get some perspective. And indeed what we see is that some core of modern computational thinking was possible even long before modern times. But the ambient technology and understanding of past centuries put definite limits on how far the thinking could go.

And of course this leads to a sobering question for us today: how much are we failing to realize from the core computational way of thinking because we do not have the ambient technology of the distant future? For me, looking at Leibniz has put this question in sharper focus. And at least one thing seems fairly clear.

In Leibniz's whole life, he basically saw less than a handful of computers, and all they did was basic arithmetic. Today there are billions of computers in the world, and they do all sorts of things. But in the future there will surely be far far more computers (made easier to create by the Principle of Computational Equivalence). And no doubt we'll get to the point where basically everything we make will explicitly be made of computers at every level. And the result is that absolutely everything will be programmable, down to atoms. Of course, biology has in a sense already achieved a restricted version of this. But we will be able to do it completely and everywhere.

At some level we can already see that this implies some merger of computational and physical processes. But just how may be as difficult for us to imagine as things like Mathematica and Wolfram|Alpha would have been for Leibniz.

Leibniz died on November 14, 1716. In 2016 that'll be 300 years ago. And it'll be a good opportunity to make sure everything we have from Leibniz has finally been gone through—and to celebrate after three centuries how many aspects of Leibniz's core vision are finally coming to fruition, albeit in ways he could never have imagined.

Benoit Mandelbrot

November 22, 2012

In nature, technology and art the most common form of regularity is repetition: a single element repeated many times, as on a tile floor. But another form is possible, in which smaller and smaller copies of a pattern are successively nested inside each other, so that the same intricate shapes appear no matter how much you "zoom in" to the whole. Fern leaves and Romanesco broccoli are two examples from nature.

One might have thought that such a simple and fundamental form of regularity would have been studied for hundreds, if not thousands, of years. But it was not. In fact, it rose to prominence only over the past 30 or so years—almost entirely through the efforts of one man, the mathematician Benoit Mandelbrot, who died in 2010 just before completing this autobiography.

Born to a Jewish family in Poland in 1924, he was the son of a dentist mother and a businessman father. One of his uncles, Szolem Mandelbrojt, was a fairly successful pure mathematician in France, and it was there that the Mandelbrots fled in 1936 as the situation for Jews in Poland worsened.

After hiding out during the war in the countryside, Mandelbrot wound up as an outstanding math student at a top technical college in Paris. At the time, French mathematics was dominated by the highly abstract Bourbaki movement, which was named for the collective pseudonym with which its members signed their work. When Mandelbrot graduated in 1947, instead of joining them, he decided to study aeronautical engineering at Caltech. He was determined, he writes, to stay away from established mathematics so that he could "feel the excitement of being the first to find a degree of order in some real, concrete, and complex area where everyone else saw a lawless mess."

Originally written as a review of Mandelbrot's posthumous autobiography, The Fractalist.

Returning to France two years later, he got involved with information theory and, from that, statistical physics and the structure of language. He ended up writing his PhD on "Games of Communication," notably investigating how the frequencies with which different words were used in a text followed a distribution known as a "power law". He did military service as a kind of science research scout, spent time at a Philips color television lab, and then began life as a "wandering scientist", partly supported by the French government.

His early destinations included Princeton (where he worked with the game-theory pioneer John von Neumann) and Geneva (with the psychologist Jean Piaget). And then, in 1958, he visited IBM, ostensibly to work on a new automated-translation project—and ended up staying for 35 years.

IBM saw its research division, in part, as a way to spread its reputation, and it allowed Mandelbrot to continue wandering and spend time as a visitor at universities like Harvard and Yale (where he eventually spent the last years of his career). His work moved from the statistics of languages to the statistics of economic systems and, by the mid-1960s, to hydrodynamics and hydrology. His typical mode of operation was to apply fairly straightforward mathematics (typically from the theory of random processes) to areas that had barely seen the light of serious mathematics before. He calls himself a "would-be Kepler of complexity", evoking (as he does continually) Johannes Kepler, the 17th-century scientist who determined the laws that describe the movement of the planets.

But in the early 1970s a mathematician friend (Mark Kac) made a crucial suggestion: Stop writing lots of papers about strangely diverse topics and instead tie them together in a book. The unifying theme could have been a technical one—in which case few people today would have probably ever heard of Benoit Mandelbrot.

But instead, perhaps just through the very act of exposition (his autobiography does not make it clear), Mandelbrot ended up doing a great piece of science and identifying a much stronger and more fundamental idea—put simply, that there are some geometric shapes, which he called "fractals", that are equally "rough" at all scales. No matter how close you look, they never get simpler, much as the section of a rocky coastline you can see at your feet looks just as jagged as the stretch you can see from space. This insight formed the core of his breakout 1975 book, *Fractals*.

Before this book, Mandelbrot's work had been decidedly numerical, with simple graphs being about as visual as most of his papers got. But between his access to computer graphics at IBM and publishers with distinctly visual orientations, his book ended up filled with striking illustrations, with his theme presented in a highly geometric way.

There was no book like it. It was a new paradigm, both in presentation and in the informal style of explanation it employed. Papers slowly started appearing (quite often with Mandelbrot as co-author) that connected it to different fields, such as biology and social science. Results were mixed and often controversial. Did the paths traced by animals or graphs of stock prices really have nested structures or follow exact power laws? Even Mandelbrot himself began to water down his message, introducing concepts like "multifractals".

But independent of science, fractals began to thrive in computer graphics, particularly in making what Mandelbrot called "forgeries" of natural phenomena—surprisingly lifelike images of trees or mountains. Some mathematicians began to investigate fractals in abstract terms, connecting them to a branch of mathematics concerned with so-called iterated maps. These had been studied in the early 1900s—notably by French mathematicians whom Mandelbrot had known as a student. But after a few results, their investigation had largely run out of steam.

Armed with computer graphics, however, Mandelbrot was able to move forward, discovering in 1979 the intricate shape known as the Mandelbrot set. The set, with its distinctive bulb-like lobes, lent itself to colorful renderings that helped fractals take hold in both the popular and scientific mind. And although I consider the Mandelbrot set in some ways a rather arbitrary mathematical object, it has been a fertile source of pure mathematical questions—as well as a striking example of how visual complexity can arise from simple rules.

In many ways, Mandelbrot's life is a heroic story of discovery. He was a great scientist, whose virtues I am keen to extol. For all his achievements, however, he could be a difficult man, as I discovered in interactions over the course of nearly 30 years. He was constantly seeking validation and constantly fighting to get his due, a theme that is already clear on the first page of his autobiography: "Let me introduce myself. A scientific warrior of sorts, and an old man now, I have written a great deal but never acquired a predictable audience."

He campaigned for the Nobel Prize in physics; later it was economics. I used to ask him why he cared so much. I pointed out that really great science—like fractals—tends to be too original for there to be prizes defined for it. But he would slough off my comments and recite some other evidence for the greatness of his achievements.

In his way, Mandelbrot paid me some great compliments. When I was in my 20s, and he in his 60s, he would ask about my scientific work: "How can so many people take someone so young so seriously?" In 2002, my book *A New Kind of Science*—in which I argued that many phenomena across science are the complex results of relatively simple, program-like rules—appeared. Mandelbrot seemed to see it as a direct threat, once declaring that "Wolfram's 'science' is not new except when it is clearly wrong; it deserves to be completely disregarded." In private, though, several mutual friends told me, he fretted that in the long view of history it would overwhelm his work.

Every time I saw him, I would explain why I thought he was wrong to worry and why, even though in the "computational universe" that my book described fractals appear in just a small corner, they will nevertheless always be fundamentally important—not least as perhaps the unique intermediate step between simple repetitive regularity and the apparent randomness of more complex computational processes.

I saw Mandelbrot quite frequently in the years before his death, and many questions he answered with the phrase, "Read my autobiography." And indeed it does answer some questions—especially about his early life—though it leaves quite a few unanswered, such as how exactly his breakthrough 1975 book came together. But knowing more about Benoit

Mandelbrot as a person helps to illuminate his work and to illustrate what it takes for great new science to be created. *The Fractalist* is a well-written tale of a scientific life, complete with first-person accounts of a surprising range of scientific greats.

Steve Jobs

October 6, 2011

I'm so sad this evening—as millions are—to hear of Steve Jobs's death. Scattered over the last quarter century, I learned much from Steve Jobs, and was proud to consider him a friend. And indeed, he contributed in various ways to all three of my major life projects so far: Mathematica, *A New Kind of Science* and Wolfram|Alpha.

I first met Steve Jobs in 1987, when he was quietly building his first NeXT computer, and I was quietly building the first version of Mathematica. A mutual friend had made the introduction, and Steve Jobs wasted no time in saying that he was planning to make the definitive computer for higher education, and he wanted Mathematica to be part of it. I don't now remember the details of our first meeting, but at the end of it, Steve gave me his business card, which tonight I found duly still sitting in my files:

Over the months after our first meeting, I had all sorts of interactions with Steve about Mathematica. Actually, it wasn't yet called Mathematica then, and one of the big topics of discussion was what it should be called. At first it had been Omega (yes, like Alpha) and later PolyMath. Steve thought those were lousy names. I gave him lists of names I'd considered, and pressed him for his suggestions. For a while he wouldn't suggest anything. But then one day he said to me: "You should call it Mathematica."

I'd actually considered that name, but rejected it. I asked Steve why he thought it was good, and he told me his theory for a name was to start from the generic term for something, then romanticize it. His favorite example at the time was Sony's Trinitron. Well, it went back and forth for a while. But in the end I agreed that, yes, Mathematica was a good name. And so it has been now for nearly 24 years.

As Mathematica was being developed, we showed it to Steve Jobs quite often. He always claimed he didn't understand the math of it (though I later learned from a good friend of mine who had known Steve in high school that Steve had definitely taken at least one calculus course). But he made all sorts of "make it simpler" suggestions about the interface and the documentation. With one slight exception, perhaps of at least curiosity interest to Mathematica aficionados: he suggested that cells in Mathematica notebook documents (and now CDFs) should be indicated not by simple vertical lines—but instead by brackets with little serifs at their ends. And as it happens, that idea opened the way to thinking of hierarchies of cells, and ultimately to many features of symbolic documents.

In June 1988 we were ready to release Mathematica. But NeXT had not yet released its computer, Steve Jobs was rarely seen in public, and speculation about what NeXT was up to had become quite intense. So when Steve Jobs agreed that he would appear at our product announcement, it was a huge thing for us.

He gave a lovely talk, discussing how he expected more and more fields to become computational, and to need the services of algorithms and of Mathematica. It was a very clean statement of a vision which has indeed worked out as he predicted. (And now it's nice when I hear through the grapevine that there are all sorts of algorithms central to the iPhone that were developed with the help of Mathematica.)

A while later, the NeXT was duly released, and a copy of Mathematica was bundled with every computer. Although the NeXT was not in its own right a commercial success, Steve's decision to bundle Mathematica turned out to be a really good idea, and was often quoted as the #1 reason people had bought NeXTs.

And as a curious footnote to history (which I learned years later), one batch of NeXTs bought for the purpose of running Mathematica went to CERN in Geneva, Switzerland—where they ended up having no less distinction than being the computers on which the web was first developed.

I used to see Steve Jobs with some regularity in those days. One time I went to see him in NeXT's swank new offices in Redwood City. I particularly wanted to talk to him about Mathematica as a computer language. He always preferred user interfaces to languages, but he was trying to be helpful. The conversation was going on, but he said he couldn't go to dinner, and actually he was quite distracted, because he was going out on a date that evening—and he hadn't been on a date for a long time. He explained that he'd just met the woman he was seeing a few days earlier, and was very nervous about his date. The Steve Jobs—so confident as a businessman and technologist—had melted away, and he was asking me—hardly a noted authority on such things—about his date.

As it turned out, the date apparently worked out—and within 18 months the woman he met became his wife, and remained so until the end.

My direct interactions with Steve Jobs decreased during the decade that I was for all practical purposes a hermit working on *A New Kind of Science*. For most of that time, though, I used a NeXT computer in almost every waking hour—and in fact my main discoveries were made on it. And when the book was finished, Steve asked for a pre-release copy, which I duly sent.

At the time, all sorts of people were telling me that I needed to put quotes on the back cover of the book. So I asked Steve Jobs if he'd give me one. Various questions came back. But eventually Steve said, "Isaac Newton didn't have back-cover quotes; why do you want them?" And that's how, at the last minute, the back cover of *A New Kind of Science* ended up with just a simple and elegant array of pictures. Another contribution from Steve Jobs, that I notice every time I look at my big book.

In my life, I have had the good fortune to interact with all sorts of talented people. To me, Steve Jobs stands out most for his clarity of thought. Over and over again he took complex situations, understood their essence, and used that understanding to make a bold definitive move, often in a completely unexpected direction.

I myself have spent much of my life—in science and in technology—trying to work in somewhat similar ways. And trying to build the very best possible things I can.

Yet looking at the practical world of technology and business there are certainly times when it has not been obvious that any of this is a good strategy. Indeed, sometimes it looks as if all that clarity, understanding, quality and new ideas aren't really the point—and that the winners are those with quite different interests.

So for me—and our company—it has been immensely inspiring to watch Steve Jobs's—and Apple's—amazing success in recent years. It validates so many of the principles that I have long believed in. And encourages me to pursue them with even greater vigor.

I think that over the years Steve Jobs appreciated the approach I've tried to take with our company. He was certainly always a great supporter. (Just tonight, for example, I was reminded of a terrific video that he sent us for the 10th-anniversary Mathematica user conference.) And he was always keen for us to work first with NeXT, and later with Apple.

I think Mathematica may hold the distinction of having been the only major software system available at launch on every single computer that Steve Jobs created since 1988. Of course, that's often led to highly secretive emergency Mathematica porting projects—culminating a couple of times in Theo Gray demoing the results in Steve Jobs's keynote speeches.

When Apple started producing the iPod and iPhone I wasn't sure how they would relate to anything we did. But after Wolfram|Alpha came out, we started realizing just how powerful it was to have computational knowledge on this new platform that Steve Jobs had created. And when the iPad was coming out, Theo Gray—at Steve Jobs's urging—insisted that we had to do something significant for it.

The result was the formation last year of Touchpress, the publication of Theo's *The Elements* iPad ebook, and now a string of other iPad ebooks. A whole new direction made possible by Steve Jobs's creation of the iPad.

It's hard to remember tonight all the ways Steve Jobs has supported and encouraged us over the years. Big things and small things. Looking at my archive I realize I'd forgotten just how many detailed problems he jumped in to solve. From the glitches in versions of NeXTSTEP, to the personal phone call not long ago to assure us that if we ported Mathematica and CDF to iOS they wouldn't be banned.

There is much that I am grateful to Steve Jobs for. But tragically, his greatest contribution to my latest life project—Wolfram|Alpha—happened just yesterday: the announcement that Wolfram|Alpha will be used in Siri on the iPhone 4S.

It is somehow a quintessential Steve Jobs move. To realize that people just want direct access to knowledge and actions on their phones. Without all the extra steps that people would usually assume have to be there.

I'm proud that we are in a position to provide an important component for that vision with Wolfram|Alpha. What's coming out now is just a beginning, and I look forward to what we will do with Apple in this direction in the future. I'm just sad that Steve Jobs will now not be part of it.

When I first met Steve Jobs nearly 25 years ago I was struck by him explaining to me that NeXT was what he "wanted to do with his thirties". At the time, I thought it was a bold thing to plan one's life in decades like that. And—particularly for those of us who spend their lives doing large projects—it's incredibly inspiring to see what Steve Jobs was able to achieve in his small number of decades, so tragically cut short today.

Thank you, Steve, for everything.

Marvin Minsky

January 26, 2016

I think it was 1979 when I first met Marvin Minsky, while I was still a teenager working on physics at Caltech. It was a weekend, and I'd arranged to see Richard Feynman to discuss some physics. But Feynman had another visitor that day as well, who didn't just want to talk about physics, but instead enthusiastically brought up one unexpected topic after another.

That afternoon we were driving through Pasadena, California—and with no apparent concern to the actual process of driving, Feynman's visitor was energetically pointing out all sorts of things an AI would have to figure if it was to be able to do the driving. I was a bit relieved when we arrived at our destination, but soon the visitor was on to another topic, talking about how brains work, and then saying that as soon as he'd finished his next book he'd be happy to let someone open up his brain and put electrodes inside, if they had a good plan to figure out how it worked.

Feynman often had eccentric visitors, but I was really wondering who this one was. It took a couple more encounters, but then I got to know that eccentric visitor as Marvin Minsky, pioneer of computation and AI—and was pleased to count him as a friend for more than three decades.

Just a few days ago I was talking about visiting Marvin—and I was so sad when I heard he died. I started reminiscing about all the ways we interacted over the years, and all the interests we shared. Every major project of my life I discussed with Marvin, from SMP, my first big software system back in 1981, through Mathematica, *A New Kind of Science*, Wolfram|Alpha and most recently the Wolfram Language.

This picture is from one of the last times I saw Marvin. His health was failing, but he was keen to talk. Having watched more than 35 years of my life, he wanted to tell me his assessment: "You really did it, Steve." Well, so did you, Marvin! (I'm always "Stephen", but somehow Americans of a certain age have a habit of calling me "Steve".)

The Marvin that I knew was a wonderful mixture of serious and quirky. About almost any subject he'd have something to say, most often quite unusual. Sometimes it'd be really interesting; sometimes it'd just be unusual. I'm reminded of a time in the early 1980s when I was visiting Boston and subletting an apartment from Marvin's daughter Margaret (who was in Japan at the time). Margaret had a large and elaborate collection of plants, and one day I noticed that some of them had developed nasty-looking spots on their leaves.

Being no expert on such things (and without the web to look anything up!), I called Marvin to ask what to do. What ensued was a long discussion about the possibility of developing microrobots that could chase mealybugs away. Fascinating though it was, at the end of it I still had to ask, "But what should I *actually* do about Margaret's plants?" Marvin replied, "Oh, I guess you'd better talk to my wife."

For many decades, Marvin was perhaps the world's greatest energy source for artificial intelligence research. He was a fount of ideas, which he fed to his long sequence of students at MIT. And though the details changed, he always kept true to his goal of figuring out how thinking works, and how to make machines do it.

Marvin the Computation Theorist

By the time I knew Marvin, he tended to talk mostly about theories where things could be figured out by what amounts to common sense, perhaps based on psychological or philosophical reasoning. But earlier in his life, Marvin had taken a different approach. His 1954 PhD thesis from Princeton was about artificial neural networks ("Theory of Neural-Analog Reinforcement Systems and Its Application to the Brain Model Problem") and it was a mathematics thesis, full of technical math. And in 1956, for example, Marvin published a paper entitled "Some Universal Elements for Finite Automata", in which he talked about how "complicated machinery can be constructed from a small number of basic elements".

This particular paper considered only essentially finite machines, based directly on specific models of artificial neural networks. But soon Marvin was looking at more general computational systems, and trying to see what they could do. In a sense, Marvin was beginning just the kind of exploration of the computational universe that years later I would also do, and eventually write *A New Kind of Science* about. And in fact, as early as 1960, Marvin came extremely close to discovering the same core phenomenon I eventually did.

In 1960, as now, Turing machines were used as a standard basic model of computation. And in his quest to understand what computation—and potentially brains—could be built from, Marvin started looking at the very simplest Turing machines (with just 2 states and 2 colors) and using a computer to find out what all 4096 of them actually do. Most he discovered just have repetitive behavior, and a few have what we'd now call nested or fractal behavior. But none do anything more complicated, and indeed Marvin based the final exercise in his classic 1967 book *Computation: Finite and Infinite Machines* on this, noting that "D. G. Bobrow and the author did this for all (2,2) machines [1961, unpublished] by a tedious reduction to thirty-odd cases (unpublishable)."

Years later, Marvin told me that after all the effort he'd spent on the (2,2) Turing machines he wasn't inclined to go further. But as I finally discovered in 1991, if one just looks at (2,3) Turing machines, then among the 3 million or so of them, there are a few that don't just show simple behavior any more—and instead generate immense complexity even from their very simple rules.

Back in the early 1960s, even though he didn't find complexity just by searching simple "naturally occurring" Turing machines, Marvin still wanted to construct the simplest one he could that would exhibit it. And through painstaking work, he came up in 1962 with a (7,4) Turing machine that he proved was universal (and so, in a sense, capable of arbitrarily complex behavior).

At the time, Marvin's (7,4) Turing machine was the simplest known universal Turing machine. And it kept that record essentially unbroken for 40 years—until I finally published a (2,5) universal Turing machine in *A New Kind of Science*. I felt a little guilty taking the record away from Marvin's machine after so long. But Marvin was very nice about it. And a few years later he enthusiastically agreed to be on the committee for a prize I put up to establish whether a (2,3) Turing machine that I had identified as the simplest possible candidate for universality was in fact universal.

It didn't take long for a proof of universality to be submitted, and Marvin got quite involved in some of the technical details of validating it, noting that perhaps we should all have known something like this was possible, given the complexity that Emil Post had observed with the simple rules of what he called a tag system—back in 1921, before Marvin was even born.

Marvin and Neural Networks

When it came to science, it sometimes seemed as if there were two Marvins. One was the Marvin trained in mathematics who could give precise proofs of theorems. The other was the Marvin who talked about big and often quirky ideas far away from anything like mathematical formalization.

I think Marvin was ultimately disappointed with what could be achieved by mathematics and formalization. In his early years he had thought that with simple artificial neural networks—and maybe things like Turing machines—it would be easy to build systems that worked like brains. But it never seemed to happen. And in 1969, with his long-time mathematician collaborator Seymour Papert, Marvin wrote a book that proved that a certain simple class of neural networks known as perceptrons couldn't (in Marvin's words) "do anything interesting".

To Marvin's later chagrin, people took the book to show that no neural network of any kind could ever do anything interesting, and research on neural networks all but stopped. But a bit like with the (2,2) Turing machines, much richer behavior was actually lurking just out of sight. It started being noticed in the 1980s, but it's only been in the last couple of years—with computers able to handle almost-brain-scale networks—that the richness of what neural networks can do has begun to become clear.

And although I don't think anyone could have known it then, we now know that the neural networks Marvin was investigating as early as 1951 were actually on a path that would ultimately lead to just the kind of impressive AI capabilities he was hoping for. It's a pity it took so long, and Marvin barely got to see it. (When we released our neural-network-based image identifier last year, I sent Marvin a pointer saying "I never thought neural networks would actually work… but…" Sadly, I never ended up talking to Marvin about it.)

Marvin and Symbolic AI

Marvin's earliest approaches to AI were through things like neural networks. But perhaps through the influence of John McCarthy, the inventor of LISP, with whom Marvin started the MIT AI Lab, Marvin began to consider more "symbolic" approaches to AI as well. And in 1961 Marvin got a student of his to write a program in LISP to do symbolic integration. Marvin told me that he wanted the program to be as "human like" as possible—so every so often it would stop and say "Give me a cookie", and the user would have to respond "A cookie".

By the standards of Mathematica or Wolfram|Alpha, the 1961 integration program was very primitive. But I'm certainly glad Marvin had it built. Because it started a sequence of projects at MIT that led to the MACSYMA system that I ended up using in the 1970s—that in many ways launched my efforts on SMP and eventually Mathematica.

Marvin himself, though, didn't go on thinking about using computers to do mathematics, but instead started working on how they might do the kind of tasks that all humans—including children—routinely do. Marvin's collaborator Seymour Papert, who had worked with developmental psychologist Jean Piaget, was interested in how children learn, and Marvin got quite involved in Seymour's project of developing a computer language for children. The result was Logo—a direct precursor of Scratch—and for a brief while in the 1970s Marvin and Seymour had a company that tried to market Logo and a hardware "turtle" to schools.

For me there was always a certain mystique around Marvin's theories about AI. In some ways they seemed like psychology, and in some ways philosophy. But occasionally there'd actually be pieces of software—or hardware—that claimed to implement them, often in ways that I didn't understand very well.

Probably the most spectacular example was the Connection Machine, developed by Marvin's student Danny Hillis and his company Thinking Machines (for which Richard Feynman and I were both consultants). It was always in the air that the Connection Machine was built to implement one of Marvin's theories about the brain, and might be seen one day as like the "transistor of artificial intelligence". But I, for example, ended up using its massively parallel architecture to implement cellular automaton models of fluids, and not anything AI-ish at all.

Marvin was always having new ideas and theories. And even as the Connection Machine was being built, he was giving me drafts of his book *The Society of Mind*, which talked about new and different approaches to AI. Ever one to do the unusual, Marvin told me he thought about writing the book in verse. But instead the book is structured a bit like so many conversations I had with Marvin: with one idea on each page, often good, but sometimes not—yet always lively.

I think Marvin viewed *The Society of Mind* as his magnum opus, and I think he was disappointed that more people didn't understand and appreciate it. It probably didn't help that the book came out in the 1980s, when AI was at its lowest ebb. But somehow I think to really appreciate what's in the book one would need Marvin there, presenting his ideas with his characteristic personal energy and responding to any objections one might have about them.

Marvin and Cellular Automata

Marvin was used to having theories about thinking that could be figured out just by thinking—a bit like the ancient philosophers had done. But Marvin was interested in everything, including physics. He wasn't an expert on the formalism of physics, though he did make contributions to physics topics (notably patenting a confocal microscope). And through his long-time friend Ed Fredkin, he had already been introduced to cellular automata in the early 1960s. He really liked the philosophy of having physics based on them—and ended up for example writing a paper entitled "Nature Abhors an Empty Vacuum" that talked about how one might in effect engineer certain features of physics from cellular automata.

Marvin didn't do terribly much with cellular automata, though in 1970 he and Fredkin used something like them in the Triadex Muse digital music synthesizer that they patented and marketed—an early precursor of cellular-automaton-based music composition.

Marvin was very supportive of my work on cellular automata and other simple programs, though I think he found my orientation towards natural science a bit alien. During the decade that I worked on *A New Kind of Science* I interacted with Marvin with some regularity. He was starting work on a book then too, about emotions, that he told me in 1992 he hoped "might reform how people think about themselves". I talked to him occasionally about his book, trying I suppose to understand the epistemological character of it (I once asked if it was a bit like Freud in this respect, and he said yes). It took 15 years for Marvin to finish what became *The Emotion Machine*. I know he had other books planned too; in 2006, for example, he told me he was working on a book on theology that was "a couple of years away"—but which sadly never saw the light of day.

Marvin in Person

It was always a pleasure to see Marvin. Often it would be at his big house in Brookline, Massachusetts. As soon as one entered, Marvin would start saying something unusual. It could be, "What would we conclude if the sun didn't set today?" Or, "You've got to come see the actual binary tree in my greenhouse." Once someone told me that Marvin could give a talk about almost anything, but if one wanted it to be good, one should ask him an interesting question just before he started, and then that'd be what he would talk about. I realized this was how to handle conversations with Marvin too: bring up a topic and then he could be counted on to say something unusual and often interesting about it.

I remember a few years ago bringing up the topic of teaching programming, and how I was hoping the Wolfram Language would be relevant to it. Marvin immediately launched into talking about how programming languages are the only ones that people are expected to learn to write before they can read. He said he'd been trying to convince Seymour Papert that the best way to teach programming was to start by showing people good code. He gave the example of teaching music by giving people *Eine kleine Nachtmusik*, and asking them to transpose it to a different rhythm and see what bugs occur. (Marvin was a long-time enthusiast of classical music.) In just this vein, one way the Wolfram Programming Lab that we launched just last week lets people learn programming is by starting with good code, and then having them modify it.

There was always a certain warmth to Marvin. He liked and supported people; he connected with all sorts of interesting people; he enjoyed telling nice stories about people. His house always seemed to buzz with activity, even as, over the years, it piled up with stuff to the point where the only free space was a tiny part of a kitchen table.

Marvin also had a great love of ideas. Ones that seemed important. Ones that were strange and unusual. But I think in the end Marvin's greatest pleasure was in connecting ideas with people. He was a hacker of ideas, but I think the ideas became meaningful to him when he used them as a way to connect with people.

I shall miss all those conversations about ideas—both ones I thought made sense and ones I thought didn't. Of course, Marvin was always a great enthusiast of cryonics, so perhaps this isn't the end of the story. But at least for now, farewell, Marvin, and thank you.

Russell Towle

October 10, 2008

A few times a year they would arrive. Email dispatches from an adventurous explorer in the world of geometry. Sometimes with subject lines like "Phenomenal discoveries!!!" Usually with images attached. And stories of how Russell Towle had just used Mathematica to discover yet another strange and wonderful geometrical object.

Then, this August, another email arrived, this time from Russell Towle's son: "...last night, my father died in a car accident."

I first heard from Russell Towle thirteen years ago, when he wrote to me suggesting that Mathematica's graphics language be extended to have primitives not just for polygons and cubes, but also for "polar zonohedra". I do not now recall, but I strongly suspect that at that time I had never heard of zonohedra. But Russell Towle's letter included some intriguing pictures, and we wrote back encouragingly.

There soon emerged more information. That Russell Towle lived in a hexagonal house of his own design, in a remote part of the Sierra Nevada mountains of California. That he was a fan of Archimedes, and had learned Greek to be able to understand his work better. And that he was not only an independent mathematician, but also a musician and an accomplished local historian.

In the years that followed, Russell Towle wrote countless Mathematica programs, published his work in *The Mathematica Journal*, created videos (*Regular Polytopes: The Movie*, *Joyfully Bitten Zonotope*, ...) and recently began publishing on The Wolfram Demonstrations Project.

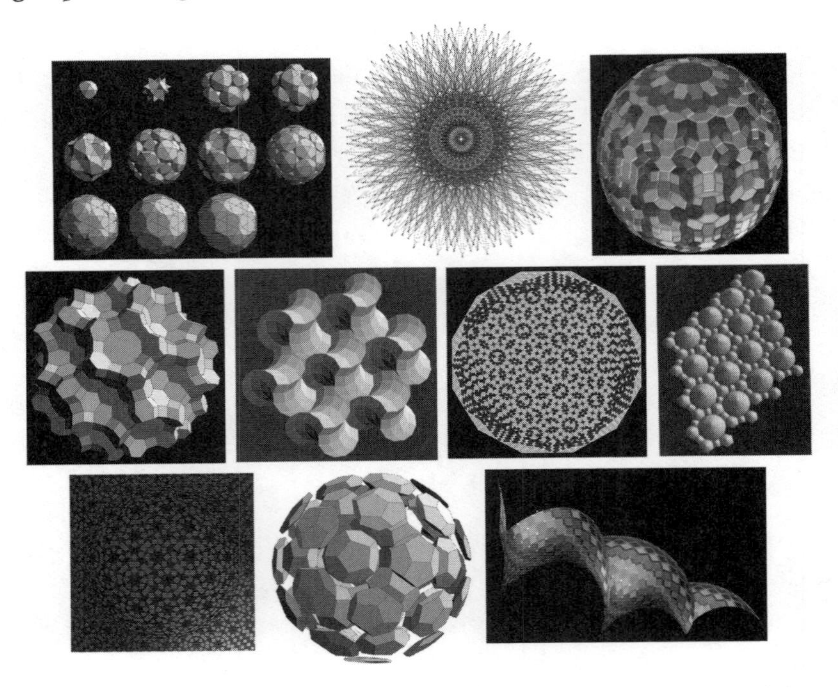

In 1996, he sent me probably the first image I had ever seen of digital elevation data rendered in Mathematica.

His last letter to me was this May, where he explained, "Along the way it occurred to me that it would be interesting to treat the zonogons of a plane tiling as pixels; an example is attached, in which I mapped a photo of a butterfly onto a patch of Penrose tiling."

Throughout the years, though, zonohedra were Russell Towle's greatest passion.

Everyone knows about the five Platonic solids, where every face and every vertex configuration has the same regular form. Then there are the 13 Archimedean solids (PolyhedronData["Archimedean"]; the cuboctahedron, icosidodecahedron, truncated cube, etc.), constructed by requiring the same configuration at each vertex, but allowing more than one kind of regular face.

The zonohedra are based on a different approach to constructing polyhedra. They start from a collection of vectors v_i at the origin, then simply consist of those regions of space corresponding to $\Sigma\ a_i\ v_i$ where the $0 < a_i < 1$.

With two vectors, this construction always gives a parallelogram. And in 3D, with three vectors, it gives a parallelepiped. And as one increases the number of vectors, one sees a lot of familiar (and not-so-familiar) polyhedra.

I'm not sure how the "known" polyhedra (included for example in PolyhedronData) are distributed in zonohedron space. That would have been a good question to ask Russell Towle. My impression is that many of the "famous" polyhedra have simple minimal representations as zonohedra. But the complete space of zonohedra contains all sorts of remarkable forms that, for whatever reason, have never arisen in the traditional historical development of polyhedra.

Russell Towle identified some particular families of zonohedra, with interesting mathematical and aesthetic properties. Zonohedra not only have all sorts of mathematical connections (they are, for example, the figures formed by projections of higher-dimensional cubes), but may, as Russell Towle suggested in his first letter to me, have practical importance as convenient parametrizations of symmetrical geometric forms.

In recent years, zonohedra have for example begun to find their way into architecture. Indeed, a 600-foot approximation to a zonohedron now graces the London skyline as the Swiss Re building ("Gherkin").

There is a certain wonderful timelessness to polyhedra. We see ancient Egyptian dice that are dodecahedra. We see polyhedra in Leonardo da Vinci's illustrations. But somehow all these polyhedra, whenever they are from, look modern.

That there is more to explore in the world of polyhedra after two thousand or more years might seem remarkable. Part of it is that we live in a time of new tools—when we can use Mathematica to explore the universe of geometrical forms. And part of it is that there have been rather few individuals who have the kind of passion, intuition and technical skill about polyhedra of a Russell Towle.

It's good that Russell Towle had the opportunity to show us a little more about the world of zonohedra; it's sad that he left us so soon, no doubt with so many fascinating kinds of zonohedra still left to discover.

Bertrand Russell & Alfred Whitehead

November 25, 2010

A hundred years ago this month the first volume of Whitehead and Russell's nearly-2000-page monumental work *Principia Mathematica* was published. A decade in the making, it contained page after page like the one below, devoted to showing how the truths of mathematics could be derived from logic.

SECTION A] THE CARDINAL NUMBER 1 351

***52·601.** $\vdash :: \alpha \epsilon 1 . \supset :. \phi(\breve{\iota}{}^{\backprime}\alpha) . \equiv : x \epsilon \alpha . \supset_x . \phi x : \equiv : (\exists x) . x \epsilon \alpha . \phi x$

Dem.

$\vdash . *52{\cdot}15 . \supset \vdash :. \mathrm{Hp} . \supset : E ! \breve{\iota}{}^{\backprime}\alpha :$ (1)

$[*30{\cdot}4] \quad \supset : x \breve{\iota} \alpha . \equiv . x = \breve{\iota}{}^{\backprime}\alpha .$

$[*52{\cdot}6] \quad \equiv . x \epsilon \alpha$ (2)

$\vdash . (1) . *30{\cdot}33 . \supset$

$\vdash :: \mathrm{Hp} . \supset :. \phi(\breve{\iota}{}^{\backprime}\alpha) . \equiv : x \breve{\iota} \alpha . \supset_x . \phi x : \equiv : (\exists x) . x \breve{\iota} \alpha . \phi x$ (3)

$\vdash . (2) . (3) . \supset \vdash . \mathrm{Prop}$

***52·602.** $\vdash :. \hat{z}(\phi z) \epsilon 1 . \supset : \psi(\imath x)(\phi x) . \equiv . \phi x \supset_x \psi x . \equiv . (\exists x) . \phi x . \psi x$ $[*52{\cdot}12 . *14{\cdot}26]$

***52·61.** $\vdash :. \alpha \epsilon 1 . \supset : \breve{\iota}{}^{\backprime}\alpha \epsilon \beta . \equiv . \alpha \subset \beta . \equiv . \exists ! (\alpha \cap \beta)$ $\left[*52{\cdot}601 \frac{x \epsilon \beta}{\phi x}\right]$

***52·62.** $\vdash :. \alpha, \beta \epsilon 1 . \supset : \alpha = \beta . \equiv . \breve{\iota}{}^{\backprime}\alpha = \breve{\iota}{}^{\backprime}\beta$

Dem.

$\vdash . *52{\cdot}601 . \supset \vdash :: \mathrm{Hp} . \supset :. \breve{\iota}{}^{\backprime}\alpha = \breve{\iota}{}^{\backprime}\beta . \equiv : x \epsilon \alpha . \supset_x . x = \breve{\iota}{}^{\backprime}\beta :$

$[*52{\cdot}6] \quad \equiv : x \epsilon \alpha . \supset_x . x \epsilon \beta :$

$[*52{\cdot}46] \quad \equiv : \alpha = \beta :: \supset \vdash . \mathrm{Prop}$

***52·63.** $\vdash : \alpha, \beta \epsilon 1 . \alpha \neq \beta . \supset . \alpha \cap \beta = \Lambda$ $[*52{\cdot}46 . \mathrm{Transp}]$

***52·64.** $\vdash : \alpha \epsilon 1 . \supset . \alpha \cap \beta \epsilon 1 \cup \iota{}^{\backprime}\Lambda$

Dem.

$\vdash . *52{\cdot}43 . \quad \supset \vdash : \mathrm{Hp} . \exists ! \alpha \cap \beta . \supset . \alpha \cap \beta \epsilon 1 :$

$[*5{\cdot}6 . *24{\cdot}54] \supset \vdash :. \mathrm{Hp} . \supset : \alpha \cap \beta = \Lambda . \vee . \alpha \cap \beta \epsilon 1 :$

$[*51{\cdot}236] \quad \supset : \alpha \cap \beta \epsilon 1 \cup \iota{}^{\backprime}\Lambda :. \supset \vdash . \mathrm{Prop}$

***52·7.** $\vdash :. \beta - \alpha \epsilon 1 . \alpha \subset \xi . \xi \subset \beta . \supset : \xi = \alpha . \vee . \xi = \beta$

Dem.

$\vdash . *22{\cdot}41 . \quad \supset \vdash : \mathrm{Hp} . \xi \subset \alpha . \supset . \xi = \alpha$ (1)

$\vdash . *24{\cdot}55 . \quad \supset \vdash : \sim(\xi \subset \alpha) . \supset . \exists ! \xi - \alpha$ (2)

$\vdash . *22{\cdot}48 . \quad \supset \vdash : \mathrm{Hp} . \quad \supset . \xi - \alpha \subset \beta - \alpha$ (3)

$\vdash . (2) . (3) . \quad \supset \vdash : \mathrm{Hp} . \sim(\xi \subset \alpha) . \supset . \exists ! \xi - \alpha . \xi - \alpha \subset \beta - \alpha$ (4)

$\vdash . *52{\cdot}1 . \quad \supset \vdash : \mathrm{Hp} . \supset . (\exists x) . \beta - \alpha = \iota{}^{\backprime}x$ (5)

$\vdash . (4) . (5) . *51{\cdot}4 . \supset \vdash : \mathrm{Hp} . \sim(\xi \subset \alpha) . \supset . \xi - \alpha = \beta - \alpha .$

$[*24{\cdot}411] \quad \supset . \xi = \beta$ (6)

$\vdash . (1) . (6) . \supset \vdash . \mathrm{Prop}$

Principia Mathematica is inspiring for the obvious effort put into it—and as someone who has spent much of his life engaged in very large intellectual projects, I feel a certain sympathy towards it.

In my own work, Mathematica shares with *Principia Mathematica* the goal of formalizing mathematics—but by building on the concept of computation, it takes a rather different approach, with a quite different outcome. And in *A New Kind of Science*, one of my objectives, also like *Principia Mathematica*, was to understand what lies beneath mathematics—though again my conclusion is quite different from *Principia Mathematica*.

Ever since Euclid there had been the notion of mathematical proof as a formal activity. But there had always been a tacit assumption that mathematics—with its numbers and geometrical figures—was still at some level talking about things in the natural world.

In the mid-1800s, however, that began to change, notably with the introduction of non-Euclidean geometries and algebras other than those of ordinary numbers. And by the end of the 1800s, there was a general movement towards thinking of mathematics as abstract formalism, independent of the natural world.

Meanwhile, ever since Aristotle, there had in a sense been another kind of formalism—logic—which was originally intended to represent specific kinds of idealized human arguments, but had gradually become assumed to represent any valid form of reasoning. For most of its history, logic had been studied, and taught, quite separately from mathematics. But in the 1800s, there began to be connections.

George Boole showed how basic logic could be formulated in algebraic terms ("Boolean algebra"). And then Gottlob Frege, working in some isolation in Germany, developed predicate logic ("for all", "there exists", etc.), and used a version of set theory to try to describe numbers and mathematics in purely logical terms.

And it was into this context that *Principia Mathematica* was born. Its two authors brought different things to the project. Alfred North Whitehead was an established Cambridge academic, who, in 1898, at the age of 38, had published *A Treatise on Universal Algebra* "to present a thorough investigation of the various systems of Symbolic Reasoning allied to ordinary Algebra". The book discussed Boolean algebra, quaternions and the theory of matrices—using them as the basis for a clean, if fairly traditional, treatment of topics in algebra and geometry.

Bertrand Russell was a decade younger. He had studied mathematics as an undergraduate in Cambridge, and by 1900, at the age of 28, he had already published books ranging in topic from German social democracy, to foundations of geometry, and the philosophy of Leibniz.

The nature of mathematics and mathematical truth was a common subject of debate among philosophers—as it had been at some level since Plato—and Russell seems to have believed that by making use of the latest advances, he could once and for all resolve the debates. In 1903, he published *The Principles of Mathematics*, volume 1 (no volume 2 was ever published)—in essence a survey, without formalism, of how standard areas of mathematics could be viewed in logical terms.

His basic concept was that by tightening up all relevant definitions using logic, it should be possible to derive every part of mathematics in a rigorous way, and thereby immediately answer questions about its nature and philosophy. But in 1901, as he tried to understand the concept of infinity in logical terms, and thought about ancient logical problems like the liar's paradox ("this statement is false"), he came across what seemed to be a fundamental inconsistency: a paradox of self reference ("Russell's paradox") about whether the set of all sets that do not contain themselves in fact contains itself.

To resolve this, Russell introduced what is often viewed as his most original contribution to mathematical logic: his theory of types—which in essence tries to distinguish between sets, sets of sets, etc. by considering them to be of different "types", and then restricts how they can be combined. I must say that I consider types to be something of a hack. And indeed I have always felt that the related idea of "data types" has very much served to hold up the long-term development of programming languages. (Mathematica, for example, gets great flexibility precisely from avoiding the use of types—even if internally it does use something like them to achieve various practical efficiencies.)

But back around 1900, as both Russell and Whitehead were trying to extend their formalizations of mathematics, they decided to launch into the project that would consume a decade of their lives and become *Principia Mathematica*.

Particularly since the work of Gottfried Leibniz in the late 1600s, there had been discussion of developing a notation for mathematics that transcended the imprecision of human language. In 1879 Gottlob Frege introduced his *Begriffsschrift* ("concept script")—which was a major advance in concepts and functionality, but had a strange two-dimensional layout that was almost impossible to read, or to print economically. And in the 1880s, Giuseppe Peano developed a cleaner and more linear notation, much of which is still in use today.

It did not help the dissemination of Peano's work that he chose to write his narrative text in a language of his own construction (based on classical Latin) called Interlingua. But still, in 1900, Russell went to a back-to-back pair of conferences in Paris about philosophy and mathematics (notable for being where Hilbert announced his problems), met Peano and became convinced that his attempt to formalize mathematics should be based on Peano's notation and approach. (Russell gave a distinctly pre-relativistic philosophical talk at the conference about absolute ordering of spatiotemporal events, while his wife Alys gave a talk about the education of women.)

The idea that mathematics could be built up from a small initial set of axioms had existed since the time of Euclid. But Russell and Whitehead wanted to have the smallest possible set, and have them be based not on ideas derived from observing the natural world, but instead on what they felt was the more solid and universal ground of logic.

With present-day experience of computers and programming it does not seem surprising that with enough "code" one should be able to start from basic concepts of logic and sets, and successfully build up numbers and the other standard constructs of mathematics. And indeed Frege, Peano and others had already started this process before 1900. But by its very weightiness, *Principia Mathematica* made the point seem both surprising and impactful.

Of course, it did not hurt the whole impression that it took until more than 80 pages into volume 2 to be able to prove (as "Proposition *110.643") that 1+1=2 (with the comment "[This] proposition is occasionally useful").

SECTION B] ARITHMETICAL SUM OF TWO CLASSES AND TWO CARDINALS 83

$*110{\cdot}643.\ \vdash . 1 +_c 1 = 2$

Dem.

$\vdash . *110{\cdot}632 . *101{\cdot}21{\cdot}28 . \supset$

$\vdash . 1 +_c 1 = \hat{\xi}\{(\exists y) . y \in \xi . \xi - \iota`y \in 1\}$

$[*54{\cdot}3] \quad = 2 . \supset \vdash . \text{Prop}$

The above proposition is occasionally useful. It is used at least three times, in *113·66 and *120·123·472.

$*110{\cdot}643.\ \vdash . 1 +_c 1 = 2$

Dem.

$\vdash . *110{\cdot}632 . *101{\cdot}21{\cdot}28 . \supset$

$\vdash . 1 +_c 1 = \hat{\xi}\{(\exists y) . y \in \xi . \xi - \iota`y \in 1\}$

$[*54{\cdot}3] \quad = 2 . \supset \vdash . \text{Prop}$

The above proposition is occasionally useful. It is used at least three times, in *113·66 and *120·123·472.

$\vdash . *110{\cdot}3 . \supset \vdash : \text{Nc}`\alpha = \text{Nc}`\beta +_c \text{Nc}`\gamma . \equiv . \text{Nc}`\alpha = \text{Nc}`(\beta + \gamma) .$

$[*100{\cdot}3{\cdot}31] \quad \supset . \alpha \text{ sm } (\beta + \gamma) .$

$[*73{\cdot}1] \quad \supset . (\exists R) . R \in 1 \to 1 . \text{D}`R = \alpha . \text{ɑ}`R = \downarrow \Lambda_\gamma``\iota``\beta \cup \Lambda_\beta \downarrow ``\iota``\gamma .$

$[*37{\cdot}15] \quad \supset . (\exists R) . R \in 1 \to 1 . \downarrow \Lambda_\gamma``\iota``\beta \subset \text{ɑ}`R . R`` \downarrow \Lambda_\gamma``\iota``\beta \subset \alpha .$

$[*110{\cdot}12 . *73{\cdot}22] \supset . (\exists \delta) . \delta \subset \alpha . \delta \text{ sm } \beta \qquad (2)$

$\vdash . (1) . (2) . \supset \vdash . \text{Prop}$

The above proof depends upon the fact that "Nc'α" and "Nc'β $+_c$ μ" are typically ambiguous, and therefore, when they are asserted to be equal, this must hold in *any* type, and therefore, in particular, in that type for which we have α ε Nc'α, *i.e.* for $\text{N}_0\text{c}`\alpha$. This is why the use of *100·3 is legitimate.

$*110{\cdot}72.\ \vdash : (\exists \delta) . \delta \text{ sm } \beta . \delta \subset \alpha . \equiv . (\exists \mu) . \mu \in \text{NC} . \text{Nc}`\alpha = \text{Nc}`\beta +_c \mu$

Dem.

$\vdash . *100{\cdot}321 . *110{\cdot}7 . \supset$

$\vdash :. \delta \text{ sm } \beta . \delta \subset \alpha . \supset : \text{Nc}`\delta = \text{Nc}`\beta : (\exists \mu) . \mu \in \text{NC} . \text{Nc}`\alpha = \text{Nc}`\delta +_c \mu :$

$[*13{\cdot}12] \quad \supset : (\exists \mu) . \mu \in \text{NC} . \text{Nc}`\alpha = \text{Nc}`\beta +_c \mu \qquad (1)$

$\vdash . (1) . *110{\cdot}71 . \supset \vdash . \text{Prop}$

I do not know if Russell and Whitehead intended *Principia Mathematica* to be readable by humans—but in the end Russell estimated years later that only perhaps six people had ever read the whole thing. To modern eyes, the use of Peano's dot notation instead of parentheses is particularly difficult. And then there is the matter of definitions.

At the end of volume 1, *Principia Mathematica* lists about 500 "definitions", each with a special notation. In many ways, these are the analogs of the built-in functions of Mathematica. But in *Principia Mathematica*, instead of being given English-based names, all these objects are assigned special notations. The first handful are not too difficult to understand. But by the second page one's seeing all sorts of strange glyphs, and I, at least, lose hope of ever being able to decipher what is written in them.

LIST OF DEFINITIONS

1·01. $p \supset q$
2·33. $p \vee q \vee r$
3·01. $p . q$
3·02. $p \supset q \supset r$
4·01. $p \equiv q$
4·02. $p \equiv q \equiv r$
4·34. $p . q . r$
9·01. $\sim\{(x) . \phi x\}$
9·011. $\sim(x) . \phi x$
9·02. $\sim\{(\exists x) . \phi x\}$
9·021. $\sim(\exists x) . \phi x$
9·03. $(x) . \phi x . \vee . p$
9·04. $p . \vee . (x) . \phi x$
9·05. $(\exists x) . \phi x . \vee . p$
9·06. $p . \vee . (\exists x) . \phi x$
9·07. $(x) . \phi x . \vee . (\exists y) . \psi y$
9·08. $(\exists y) . \psi y . \vee . (x) . \phi x$
10·01. $(\exists x) . \phi x$
10·02. $\phi x \supset_x \psi x$
10·03. $\phi x \equiv_x \psi x$
11·01. $(x, y) . \phi(x, y)$
11·02. $(x, y, z) . \phi(x, y, z)$
11·03. $(\exists x, y) . \phi(x, y)$
11·04. $(\exists x, y, z) . \phi(x, y, z)$
11·05. $\phi(x, y) . \supset_{x,y} . \psi(x, y)$
11·06. $\phi(x, y) . \equiv_{x,y} . \psi(x, y)$
13·01. $x = y$
13·02. $x \neq y$
13·03. $x = y = z$
14·01. $[(\iota x)(\phi x)] . \psi(\iota x)(\phi x)$
14·02. $E!(\iota x)(\phi x)$
14·03. $[(\iota x)(\phi x), (\iota x)(\psi x)] . f\{(\iota x)(\phi x), (\iota x)(\psi x)\}$
14·04. $[(\iota x)(\psi x)] . f\{(\iota x)(\phi x), (\iota x)(\psi x)\}$
20·01. $f\{\hat{z}(\psi z)\}$
20·02. $x \epsilon (\phi ! \hat{z})$
20·03. Cls
20·04. $x, y \epsilon \alpha$
20·05. $x, y, z \epsilon \alpha$
20·06. $x \sim \epsilon \alpha$
20·07. $(\alpha) . f\alpha$
20·071. $(\exists \alpha) . f\alpha$
20·072. $[(\iota\alpha)(\phi\alpha)] . f(\iota\alpha)(\phi\alpha)$
20·08. $f\{\hat{\alpha}(\psi\alpha)\}$
20·081. $\alpha \epsilon \psi ! \hat{\alpha}$
21·01. $f\{\hat{x}\hat{y}\psi(x, y)\}$
21·02. $a\{\phi ! (\hat{x}, \hat{y})\} b$
21·03. Rel
21·07. $(R) . fR$
21·071. $(\exists R) . fR$
21·072. $[(\iota R)(\phi R)] . f(\iota R)(\phi R)$
21·08. $f\{\hat{R}\hat{S}\psi(R, S)\}$
21·081. $P\{\phi ! (\hat{R}, \hat{S})\} Q$
21·082. $f\{\hat{R}(\psi R)\}$
21·083. $R \epsilon \phi ! \hat{R}$
22·01. $\alpha \subset \beta$
22·02. $\alpha \cap \beta$

668 LIST OF DEFINITIONS

22·03. $\alpha \cup \beta$
22·04. $-\alpha$
22·05. $\alpha - \beta$
22·53. $\alpha \cap \beta \cap \gamma$
22·71. $\alpha \cup \beta \cup \gamma$
23·01. $R \Subset S$
23·02. $R \mathbin{\dot\cap} S$
23·03. $R \mathbin{\dot\cup} S$
23·04. $\dot{-} R$
23·05. $R \dot{-} S$
23·53. $R \mathbin{\dot\cap} S \mathbin{\dot\cap} T$
23·71. $R \mathbin{\dot\cup} S \mathbin{\dot\cup} T$
24·01. V
24·02. Λ
24·03. $\exists ! \alpha$
25·01. $\dot{V}$
25·02. $\dot{\Lambda}$
25·03. $\dot{\exists} ! R$
30·01. $R‘y$
30·02. $R‘S‘y$
31·01. Cnv
31·02. $\breve{P}$
32·01. $\overrightarrow{R}$
32·02. $\overleftarrow{R}$
32·03. sg
32·04. gs
33·01. D
33·02. Ɑ
33·03. C
33·04. F
34·01. $R \mid S$
34·02. R^2
34·03. R^3
35·01. $\alpha \upharpoonleft R$
35·02. $R \upharpoonright \beta$
35·03. $\alpha \upharpoonleft R \upharpoonright \beta$
35·04. $\alpha \uparrow \beta$
35·05. $R‘x \uparrow \beta$
35·24. $\alpha \upharpoonleft R \mid S$
35·25. $S \mid R \upharpoonright \beta$
36·01. $P \restriction \alpha$
37·01. $R‘‘\beta$
37·02. R_ϵ
37·03. $\breve{R}_\epsilon$
37·04. $R‘‘‘\kappa$
37·05. $E!! R‘‘\beta$
38·01. x♀
38·02. ♀y
38·03. α♀y
40·01. $p‘\kappa$
40·02. $s‘\kappa$
41·01. $\dot{p}‘\lambda$
41·02. $\dot{s}‘\lambda$
43·01. $R \parallel S$
50·01. I
50·02. J
51·01. ι
52·01. 1
54·01. 0
54·02. 2
55·01. $x \downarrow y$
55·02. $R‘x \downarrow y$
56·01. $\dot{2}$
56·02. 2_r

Beyond these notational issues, there is a much more fundamental difference between the formalization of mathematics in *Principia Mathematica* and in Mathematica. For in *Principia Mathematica* the objective is to exhibit true theorems of mathematics, and to represent the processes involved in proving them. But in Mathematica, the objective is instead to compute: to take mathematical expressions, and evaluate them.

(These differences in objective lead to many differences in character. For example, *Principia Mathematica* is constantly trying to give constraints that indirectly specify whatever structure it wants to talk about. In Mathematica, the whole idea is to have explicit symbolic structures that can then be computed with.)

In the hundred years since *Principia Mathematica*, there has been slow progress in presenting theorems of mathematics in formal ways. But the idea of mathematical computation has taken off spectacularly—and has transformed the use of mathematics, and many areas of its development.

But what about the conceptual purposes of *Principia Mathematica*? Russell explained in the introduction to his *The Principles of Mathematics* that he intended to "reduce the whole of [the propositions of mathematics] to certain fundamental notions of logic." Indeed, he even made what he considered to be a very general definition of "pure mathematics" as all true logical statements that contain only variables like p and q, and not literals like "the city of New York". (Applied mathematics, he suggested, would come from replacing the variables by literals.)

But why start from logic? I think Russell just assumed that logic was the most fundamental possible thing—the ultimate incontrovertible representation for all formal processes. Traditional mathematical constructs—like numbers and space—he imagined were associated with the particulars of our world. But logic, he imagined, was a kind of "pure thought", and something more general, and wholly independent of the particulars of our world.

In my own work leading up to *A New Kind of Science*, I started by studying the natural world, yet found myself increasingly being led to generalize beyond traditional mathematical constructs. But I did not wind up with logic. Instead, I began to consider all possible kinds of rules—or as I have tended to describe it (making use of modern experience), the computational universe of all possible programs.

Some of these programs describe parts of the natural world. Some give us interesting fodder for technology. And some correspond to traditional formal systems like logic and mathematics.

One thing to do is to look at the space of all possible axiom systems. There are some technical issues about modern equational systems compared to implicational systems of the kind considered in *Principia Mathematica*. But the essential result is that dotted around the space of all possible axiom systems are the particular axiom systems that have historically arisen in the development of mathematics and related fields.

Is logic somehow special? I think not.

In *Principia Mathematica*, Russell and Whitehead originally defined logic using a fairly complicated traditional set of axioms. In the second edition of the book, they made a point of noting that by writing everything in terms of NAND (Sheffer stroke) rather than AND, OR and NOT, it is possible to use a much simpler axiom system.

In 2000, by doing a search of the space of possible axiom systems, I was able to find the very simplest (equational) axiom system for standard propositional logic: just the single axiom $((a.b).c).(a.((a.c).a)) = c$. And from this result, we can tell where logic lies in the space of possible formal systems: in a natural enumeration of axiom systems in order of size, it is about the 50,000th formal system that one would encounter.

A few other traditional areas of mathematics—like group theory—occur in a comparable place. But most require much larger axiom systems. And in the end the picture seems very different from the one Russell and Whitehead imagined. It is not that logic—as conceived by human thought—is at the root of everything. Instead, there are a multitude of possible formal systems, some picked by the natural world, some picked by historical developments in mathematics, but most out there uninvestigated.

In writing *Principia Mathematica*, one of Russell's principal objectives was to give evidence that all of mathematics really could be derived from logic. And indeed the very heft of the book gave immediate support to this idea, and gave such credibility to logic (and to Russell) that Russell was able to spend much of the rest of his long life confidently presenting logic as a successful way to address moral, social and political issues.

Of course, in 1931 Kurt Gödel showed that no finite system—logic or anything else—can be used to derive all of mathematics. And indeed the very title of his paper refers to the incompleteness of none other than the formal system of *Principia Mathematica*. By this time, however, both Russell and Whitehead had moved on to other pursuits, and neither returned to address the implications of Gödel's theorem for their project.

So can one say that the idea of logic somehow underlying mathematics is wrong? At a conceptual level, I think so. But in a strange twist of history, logic is currently precisely what is actually used to implement mathematics.

For inside all current computers are circuits consisting of millions of logic gates—each typically performing a NAND operation. And so, for example, when Mathematica runs on a computer and implements the operations of mathematics, it does so precisely by marshalling the logic operations in the hardware of the computer. (To be clear, the logic implemented by computers is basic, propositional logic—not the more elaborate predicate logic, combined with set theory, that *Principia Mathematica* ultimately uses.)

We know from computational universality—and more precisely from the Principle of Computational Equivalence—that things do not have to work this way, and that there are many very different bases for computation that could be used. And indeed, as computers move to a molecular scale, standard logic will most likely no longer be the most convenient basis to use.

But so why is logic used in today's computers? I suspect it actually has quite a bit to do with none other than *Principia Mathematica*. For historically *Principia Mathematica* did much to promote the importance and primacy of logic, and the glow that it left is in many ways still with us today. It is just that we now understand that logic is just one possible basis for what we can do—not the only conceivable one.

(People familiar with technical aspects of logic may protest that the notion of "truth" is somehow intimately tied to traditional logic. I think this is a matter of definition, but in any case, what has become clear for mathematics is that it is vastly more important to compute answers than merely to state truths.)

A hundred years after *Principia Mathematica* there is still much that we do not understand even about basic questions in the foundations of mathematics. And it is humbling to wonder what progress could be made over the next hundred years.

When we look at *Principia Mathematica*, it emphasizes exhibiting particular truths of mathematics that its authors derived. But today Mathematica in effect every day automatically delivers millions of truths of mathematics, made to order for a multitude of particular purposes.

Yet it is still the case that it operates with just a few formal systems that happen to have been studied in mathematics or elsewhere. And even in *A New Kind of Science*, I concentrated on particular programs or systems that for one reason or another I thought were interesting.

In the future, however, I suspect that there will be another level of automation. Probably it will take much less than a hundred years, but in time it will become commonplace not just to make computations to order, but to make to order the very systems on which those computations are based—in effect in the blink of an eye inventing and developing something like a whole *Principia Mathematica* to respond to some particular purpose.

Richard Crandall

December 30, 2012

Richard Crandall liked to call himself a "computationalist", for though he was trained in physics (and served for many years as a physics professor at Reed College), computation was at the center of his life. He used it in physics, in engineering, in mathematics, in biology... and in technology. He was a pioneer in experimental mathematics, and was associated for many years with Apple and with Steve Jobs, and was proud of having invented "at least 5 algorithms used in the iPhone". He was also an extremely early adopter of Mathematica, and a well-known figure in the Mathematica community. And when he died just before Christmas at the age of 64 he was hard at work on his latest, rather different project: an "intellectual biography" of Steve Jobs that I had suggested he call "Scientist to Mr. Jobs".

I first met Richard Crandall in 1987, when I was developing Mathematica, and he was Chief Scientist at Steve Jobs's company NeXT. Richard had pioneered using Pascal on Macintoshes to teach scientific computing. But as soon as he saw Mathematica, he immediately adopted it, and for a quarter of a century used it to produce a wonderful range of discoveries and inventions.

He also contributed greatly to Mathematica and its usage. Indeed, even before Mathematica 1.0 in 1988, he insisted on visiting our company to contribute his expertise in numerical evaluation of special functions (his favorites were polylogarithms and zeta-like functions). And then, after the NeXT computer was released, he wrote what may have been the first-ever Mathematica-based app: a "supercalculator" named Gourmet that he said "eats other calculators for breakfast". A couple of years later he wrote a book entitled *Mathematica for the Sciences*, that pioneered the use of Mathematica programs as a form of exposition.

Over the years, I interacted with Richard about a great many things. Usually it would start with a "call me" message. And I would get on the phone, never knowing what to expect. And Richard would be talking about his latest result in number theory. Or the latest Apple GPU. Or his models of flu epidemiology. Or the importance of running Mathematica on iOS. Or a new way to multiply very long integers. Or his latest achievements in image processing. Or a way to reconstruct fractal brain geometries.

Richard made contributions—from highly theoretical to highly practical—to a wide range of fields. He was always a little too original to be in the mainstream, with the result that there are few fields where he is widely known. In recent years, however, he was beginning to be recognized for his pioneering work in experimental mathematics, particularly as applied to primes and functions related to them. But he always knew that his work with the greatest immediate significance for the world at large was what he did for Apple behind closed doors.

Richard was born in Ann Arbor, Michigan, in 1947. His father was an actuary who became a sought-after expert witness on complex corporate insurance-fraud cases, and who, Richard told me, taught him "an absolute lack of fear of large numbers". Richard grew up in Los Angeles, studying first at Caltech (where he encountered Richard Feynman), then at Reed College in Oregon. From there he went to MIT, where he studied the mathematical physics of high-energy particle scattering (Regge theory), and got his PhD in 1973. On the side he became an electronics entrepreneur, working particularly on security systems, and inventing (and patenting) a new type of operational amplifier and a new form of alarm system. After his PhD these efforts led him to New York City, where he designed a computerized fire safety and energy control system used in skyscrapers. As a hobby he worked on quantum physics and number theory—and after moving back to Oregon to work for an electronics company there, he was hired in 1978 at Reed College as a physics professor.

Steve Jobs had ended his short stay at Reed some years earlier, but through his effort to get Reed computerized, Richard got connected to him, and began a relationship that would last the rest of Steve's life. I don't know even a fraction of what Richard worked on for NeXT and Apple. For a while he was Apple's Chief Cryptographer—notably inventing a fast form of elliptic curve encryption. And later on, he was also involved in compression, image processing, touch detection, and many other things.

Through most of this, Richard continued as a practicing physics professor. Early on, he won awards for creating minimal physics experiments ("measure the speed of light on a tabletop with $10 of equipment"). By the mid-1980s, he began to concentrate on using computers for teaching—and increasingly for research. One particular direction that Richard had pursued for many years was to use computers to study properties of numbers, and for example search for primes of particular types. And particularly once he had Mathematica, he got involved in more and more sophisticated number-theoretical mathematics, particularly around primes, among other things co-authoring the (Mathematica-assisted) definitive textbook *Prime Numbers: A Computational Perspective*.

He invented faster methods for doing arithmetic with very long integers, that were instrumental, for example, in early crowdsourced prime discoveries, and that are in fact used today in modified form in Mathematica. And by doing experimental mathematics with Mathematica he discovered a wonderful collection of zeta-function-related results and identities worthy of Ramanujan. He was particularly proud of his algorithms for the fast evaluation of various zeta-like functions (notably polylogarithms and Madelung sums), and indeed earlier this year he sent me the culmination of his 20 years of work on the subject, in the form of a paper dedicated to Jerry Keiper, the founder of the numerics group at Wolfram Research, who died in an accident in 1995 but with whom Richard had worked at length.

Richard was always keen on presentation, albeit in his own somewhat unique way. Through his "industrial algorithm" company Perfectly Scientific, he published a poster of the digits of every new Mersenne prime that was discovered. The price of the poster increased with the number of digits, and for convenience his company also sold a watchmaker's loupe to allow people to read the digits on the posters.

Richard always had a certain charming personal ponderousness to him, his conversation peppered with phrases like "let me commend to your attention". And indeed as I write this, I find a classic example of over-the-top Richardness in the opening to his *Mathematica for the Sciences*: "It has been said that the evolution of humankind took a substantial, discontinuous swerve about the time when our forepaws left the ground. Once in the air, our hands were free for 'other things'. Toolmaking....", and eventually, as he explains after his "admittedly conjectural rambling", computers and Mathematica...

Richard regularly visited Steve Jobs and his family, with his last visit being just a few days before Steve died. He was always deeply impressed by Steve, and frustrated that he felt people didn't understand the strength of Steve's intellect. He was disappointed by Walter Isaacson's highly successful biography of Steve, and had embarked on writing his own "intellectual biography" of Steve. He had years of interesting personal anecdotes about Steve and his interactions with him, but he was adamant that his book should tell "the real story", about ideas and technology, and should at all costs avoid what he at least considered "gossip". At first, he was going to try to take himself completely out of the story, but I think I successfully convinced him that with his unique role as "scientist to Steve Jobs", he had no choice but to be in the story, and indeed to tell his own story along the way.

Richard was in many ways a rather solitary individual. But he always liked talking about his now-15-year-old daughter, whom he would invariably refer to rather formally as "Ellen Crandall". He had theories about many things, including child rearing, and considered one of his signature quotes to be, "The most efficient way to raise an atheist kid is to have a priest for a father." And indeed as part of the last exchange I had with him just a few weeks before he died, he marveled that his daughter from a "pure blank, white start" ... "has suddenly taken up filling giant white poster boards with minutely detailed drawing".

While his overall health was not perfect, Richard was in many ways still in the prime of his life. He had ambitious plans for the future, in mathematics, in science and in technology, not to mention in writing his biography of Steve Jobs. But a few weeks ago, he suddenly fell ill, and within ten days he died. A life cut off far too soon. But a unique life in which much was invented that would likely never have existed otherwise.

I shall miss Richard's flow of wonderfully eccentric ideas, as well as the mysterious "call me" messages, and of late the practically monthly encouragement speech about the importance of having Mathematica on the iPhone. (I'm so sorry, Richard, that we didn't get it done in time.)

Richard was always imagining what might be possible, then in his unique way doggedly trying to build towards it. Around the world at any time of day or night millions of people are using their iPhones. And unknown to them, somewhere inside, algorithms are running that one can imagine represent a little piece of the soul of that interesting and creative human being named Richard Crandall, now cast in the form of code.

Srinivasa Ramanujan

April 27, 2016

They used to come by physical mail. Now it's usually email. From around the world, I have for many years received a steady trickle of messages that make bold claims—about prime numbers, relativity theory, AI, consciousness or a host of other things—but give little or no backup for what they say. I'm always so busy with my own ideas and projects that I invariably put off looking at these messages. But in the end I try to at least skim them—in large part because I remember the story of Ramanujan.

On about January 31, 1913 a mathematician named G. H. Hardy in Cambridge, England received a package of papers with a cover letter that began: "Dear Sir, I beg to introduce myself to you as a clerk in the Accounts Department of the Port Trust Office at Madras on a salary of only £20 per annum. I am now about 23 years of age...." and went on to say that its author had made "startling" progress on a theory of divergent series in mathematics, and had all but solved the longstanding problem of the distribution of prime numbers. The cover letter ended: "Being poor, if you are convinced that there is anything of value I would like to have my theorems published.... Being inexperienced I would very highly value any advice you give me. Requesting to be excused for the trouble I give you. I remain, Dear Sir, Yours truly, S. Ramanujan".

What followed were at least 11 pages of technical results from a range of areas of mathematics (at least 2 of the pages have now been lost). There are a few things that on first sight might seem absurd, like that the sum of all positive integers can be thought of as being equal to –1/12:

$$1+2+3+4+5+6+\ldots=-\frac{1}{12}$$

$$1+2+3+4+\&c = -\frac{1}{12}$$

$$1^3+2^3+3^3+\&c = \frac{1}{120}.$$

Then there are statements that suggest a kind of experimental approach to mathematics:

I have observed that $f(e^{2\pi x})$ is of such a nature that its value is very small when x lies between 0 and 3 (its value is less than a few hundreds when $x = 3$) and rapidly increases when x is greater than 3.

But some things get more exotic, with pages of formulas like this:

V Theorems on summations of series; e.g.

(1) $\frac{1}{1^3}\cdot\frac{1}{2} + \frac{1}{2^3}\cdot\frac{1}{2^2} + \frac{1}{3^3}\cdot\frac{1}{2^3} + \frac{1}{4^3}\cdot\frac{1}{2^4} + \&c$

$= \frac{1}{6}(\log 2)^3 - \frac{\pi^2}{12}\log 2 + \left(\frac{1}{1^3} + \frac{1}{3^3} + \frac{1}{5^3} + \&c\right).$

(2) $1 + 9\cdot\left(\frac{1}{4}\right)^4 + 17\cdot\left(\frac{1\cdot5}{4\cdot8}\right)^4 + 25\left(\frac{1\cdot5\cdot9}{4\cdot8\cdot12}\right)^4 + \&c = \frac{2\sqrt{2}}{\sqrt{\pi}\cdot\left\{\Gamma\left(\frac{3}{4}\right)\right\}^2}$

(3) $1 - 5\cdot\left(\frac{1}{2}\right)^3 + 9\cdot\left(\frac{1\cdot3}{2\cdot4}\right)^3 - \&c = \frac{2}{\pi}.$

(4) $\frac{1^{13}}{e^{2\pi}-1} + \frac{2^{13}}{e^{4\pi}-1} + \frac{3^{13}}{e^{6\pi}-1} + \&c = \frac{1}{24}.$

(5) $\frac{\coth\pi}{1^7} + \frac{\coth 2\pi}{2^7} + \frac{\coth 3\pi}{3^7} + \&c = \frac{19\pi^7}{56700}.$

(6) $\frac{1}{1^5\cosh\frac{\pi}{2}} - \frac{1}{3^5\cosh\frac{3\pi}{2}} + \frac{1}{5^5\cosh\frac{5\pi}{2}} - \&c = \frac{\pi^5}{768}.$

(7) $\frac{1}{(1^2+2^2)(\sinh 3\pi - \sinh\pi)} + \frac{1}{(2^2+3^2)(\sinh 5\pi - \sinh\pi)}$

$+ \frac{1}{(3^2+4^2)(\sinh 7\pi - \sinh\pi)} + \&c$

$= \left(\frac{1}{\pi} + \coth\pi - \frac{\pi}{2}\tanh^2\frac{\pi}{2}\right) \Big/ 2\sinh\pi.$

What are these? Where do they come from? Are they even correct?

The concepts are familiar from college-level calculus. But these are not just complicated college-level calculus exercises. Instead, when one looks closely, each one has something more exotic and surprising going on—and seems to involve a quite different level of mathematics.

Today we can use Mathematica or Wolfram|Alpha to check the results—at least numerically. And sometimes we can even just type in the question and immediately get out the answer:

In[1]:= $\sum_{n}^{\infty} \frac{\text{Coth}[n\,\pi]}{n^7}$

Out[1]= $\frac{19\,\pi^7}{56\,700}$

In[2]:= $\prod_{k}^{\infty}\left(1+\left(\frac{2\,n+1}{n+k}\right)^3\right)$

Out[2]= $\frac{\text{Cos}\left[\frac{1}{2}\sqrt{3}\,\sqrt{-(1+2\,n)^2}\,\pi\right]\text{Gamma}[1+n]^3}{\pi\,\text{Gamma}[2+3\,n]}$

And the first surprise—just as G. H. Hardy discovered back in 1913—is that, yes, the formulas are essentially all correct. But what kind of person would have made them? And how? And are they all part of some bigger picture—or in a sense just scattered random facts of mathematics?

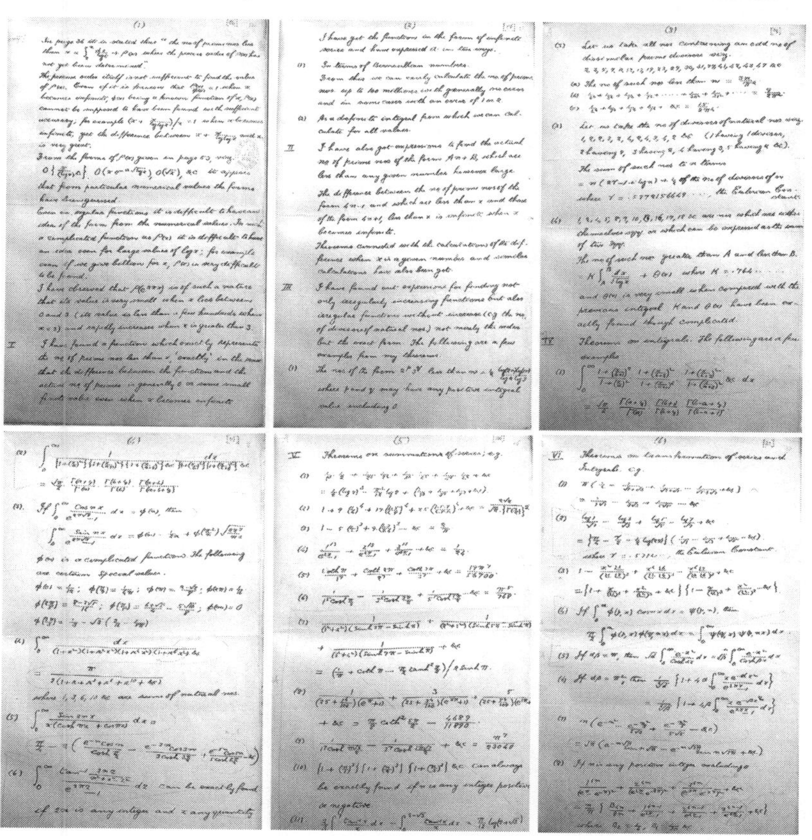

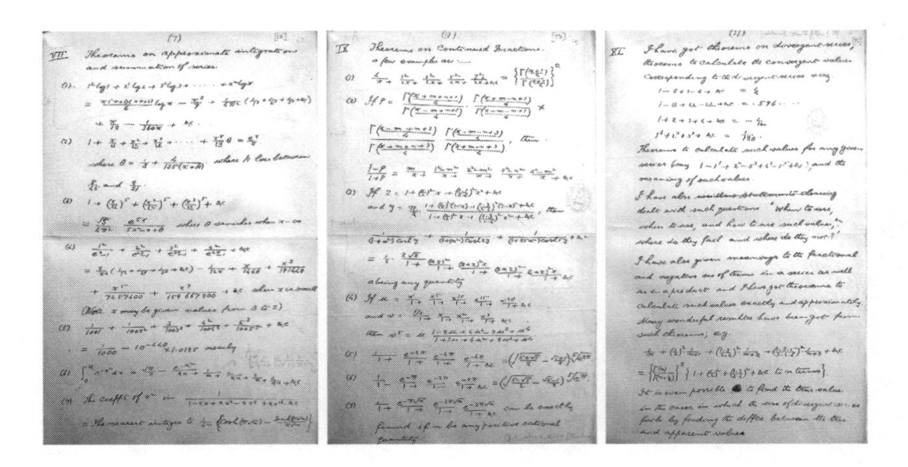

VII. Theorems on approximate integration and summation of series.

IX. Theorems on Continued Fractions. a few examples are ...

XI. I have got theorems on divergent series, theorems to calculate the convergent values corresponding to the divergent series viz.

$1 - 2 + 3 - 4 + \&c = \frac{1}{4}$

$1 - 1! + 2! - 3! + \&c = .596 \ldots$

$1 + 2 + 3 + 4 + \&c = -\frac{1}{12}$

$1^3 + 2^3 + 3^3 + \&c = \frac{1}{120}$.

Theorems to calculate such values for any given series (say $1 - 1^1 + 2^2 - 3^3 + 4^4 - 5^5 + \&c$), and the meaning of such values.

I have also dealt with such questions "When to use, where to use, and how to use such values, where do they fail and where do they not?"

I have also given meanings to the fractional and negative no. of terms in a series as well as in a product and I have got theorems to calculate such values exactly and approximately. Many wonderful results have been got from such theorems; e.g.

It is even possible to find the true value in the cases in which the sum of divergent series fails by finding the diff^ce between the true and apparent values.

The Beginning of the Story

Needless to say, there's a human story behind this: the remarkable story of Srinivasa Ramanujan.

He was born in a smallish town in India on December 22, 1887 (which made him not "about 23", but actually 25, when he wrote his letter to Hardy). His family was of the Brahmin (priests, teachers, ...) caste but of modest means. The British colonial rulers of India had put in place a very structured system of schools, and by age 10 Ramanujan stood out by scoring top in his district in the standard exams. He also was known as having an exceptional memory, and being able to recite digits of numbers like pi as well as things like roots of Sanskrit words. When he graduated from high school at age 17 he was recognized for his mathematical prowess, and given a scholarship for college.

While in high school Ramanujan had started studying mathematics on his own—and doing his own research (notably on the numerical evaluation of Euler's constant, and on properties of the Bernoulli numbers). He was fortunate at age 16 (in those days long before the web!) to get a copy of a remarkably good and comprehensive (at least as of 1886) 1055-page summary of high-end undergraduate mathematics, organized in the form of results numbered up to 6165. The book was written by a tutor for the ultra-competitive Mathematical Tripos exams in Cambridge—and its terse "just the facts" format was very similar to the one Ramanujan used in his letter to Hardy.

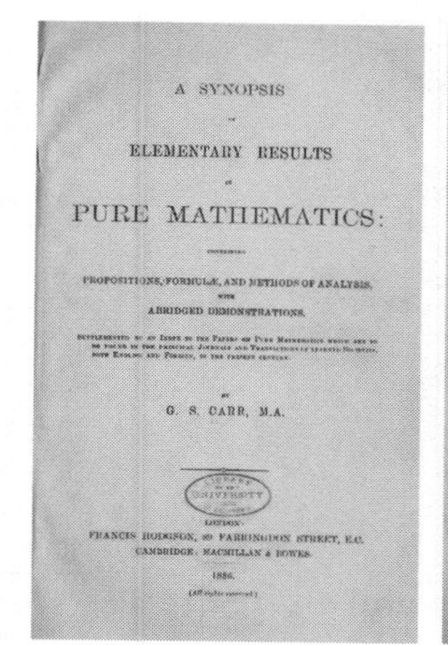

A SYNOPSIS

OF

ELEMENTARY RESULTS

IN

PURE MATHEMATICS:

CONTAINING

PROPOSITIONS, FORMULÆ, AND METHODS OF ANALYSIS,

WITH

ABRIDGED DEMONSTRATIONS.

BY

G. S. CARR, M.A.

LONDON:

FRANCIS HODGSON, 89 FARRINGDON STREET, E.C.

CAMBRIDGE: MACMILLAN & BOWES.

1886.

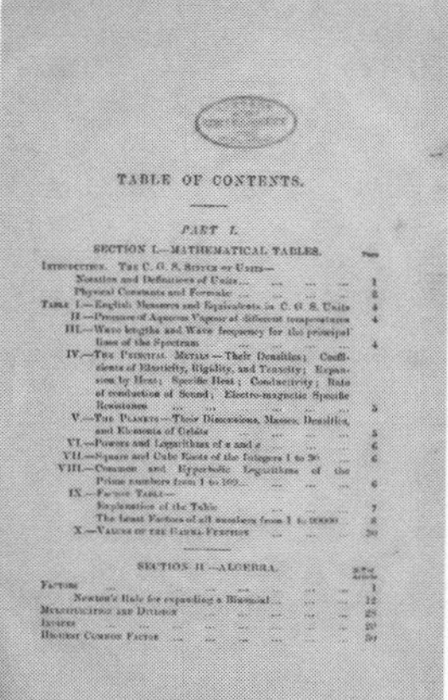

TABLE OF CONTENTS.

PART I.

SECTION I.—MATHEMATICAL TABLES.

SECTION II.—ALGEBRA.

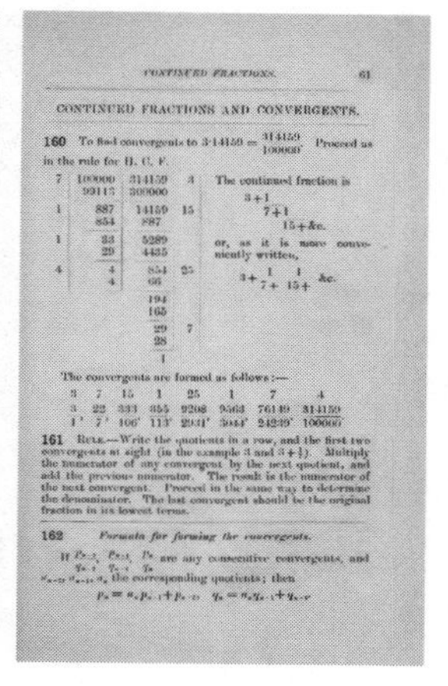

CONTINUED FRACTIONS. 61

CONTINUED FRACTIONS AND CONVERGENTS.

160 To find convergents to 3·14159 = $\frac{314159}{100000}$. Proceed as in the rule for H. C. F.

The continued fraction is

$$3+\frac{1}{7+\frac{1}{15+\&c.}}$$

or, as it is more conveniently written,

$$3+\frac{1}{7+}\frac{1}{15+}\ \&c.$$

The convergents are formed as follows:—

3	7	15	1	25	1	7	4
$\frac{3}{1}$	$\frac{22}{7}$	$\frac{333}{106}$	$\frac{355}{113}$	$\frac{9208}{2931}$	$\frac{9563}{3044}$	$\frac{76149}{24239}$	$\frac{314159}{100000}$

161 RULE.—Write the quotients in a row, and the first two convergents at sight (in the example 3 and 3 + $\frac{1}{7}$). Multiply the numerator of any convergent by the next quotient, and add the previous numerator. The result is the numerator of the next convergent. Proceed in the same way to determine the denominator. The last convergent should be the original fraction in its lowest terms.

162 *Formula for forming the convergents.*

$$p_n = a_n p_{n-1} + p_{n-2} \qquad q_n = a_n q_{n-1} + q_{n-2}$$

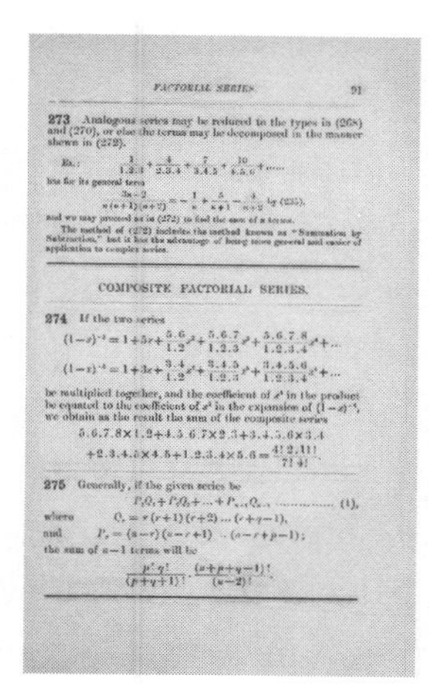

FACTORIAL SERIES. 91

273 Analogous series may be reduced to the types in (268) and (270), or else the terms may be decomposed in the manner shewn in (272).

The method of (272) includes the method known as "Summation by Subtraction," but it has the advantage of being more general and easier of application to complex series.

COMPOSITE FACTORIAL SERIES.

274 If the two series

$$(1-x)^{-5} = 1+5x+\frac{5.6}{1.2}x^2+\frac{5.6.7}{1.2.3}x^3+\frac{5.6.7.8}{1.2.3.4}x^4+\ldots$$

$$(1-x)^{-3} = 1+3x+\frac{3.4}{1.2}x^2+\frac{3.4.5}{1.2.3}x^3+\frac{3.4.5.6}{1.2.3.4}x^4+\ldots$$

be multiplied together, and the coefficient of x^4 in the product be equated to the coefficient of x^4 in the expansion of $(1-x)^{-8}$, we obtain as the result the sum of the composite series

$$5.6.7.8\times1.2+4.5.6.7\times2.3+3.4.5.6\times3.4$$
$$+2.3.4.5\times4.5+1.2.3.4\times5.6 = \frac{4!\,2!\,11!}{7!\,4!}.$$

275 Generally, if the given series be

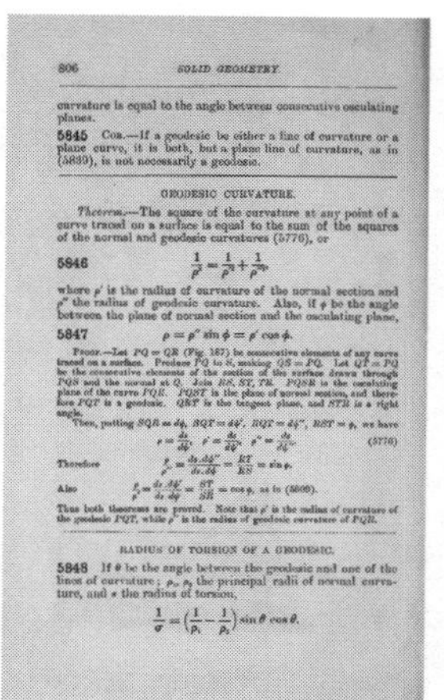

806 SOLID GEOMETRY.

curvature is equal to the angle between consecutive osculating planes.

5845 COR.—If a geodesic be either a line of curvature or a plane curve, it is both, but a plane line of curvature, as in (5839), is not necessarily a geodesic.

GEODESIC CURVATURE.

Theorem.—The square of the curvature at any point of a curve traced on a surface is equal to the sum of the squares of the normal and geodesic curvatures (5776), or

5846 $$\frac{1}{\rho^2} = \frac{1}{\rho'^2}+\frac{1}{\rho''^2},$$

where ρ' is the radius of curvature of the normal section and ρ'' the radius of geodesic curvature. Also, if ϕ be the angle between the plane of normal section and the osculating plane,

5847 $$\rho = \rho'' \sin\phi = \rho' \cos\phi.$$

Thus both theorems are proved.

RADIUS OF TORSION OF A GEODESIC.

5848 If θ be the angle between the geodesic and one of the lines of curvature; ρ_1, ρ_2 the principal radii of normal curvature, and σ the radius of torsion,

$$\frac{1}{\sigma} = \left(\frac{1}{\rho_1}-\frac{1}{\rho_2}\right)\sin\theta\cos\theta.$$

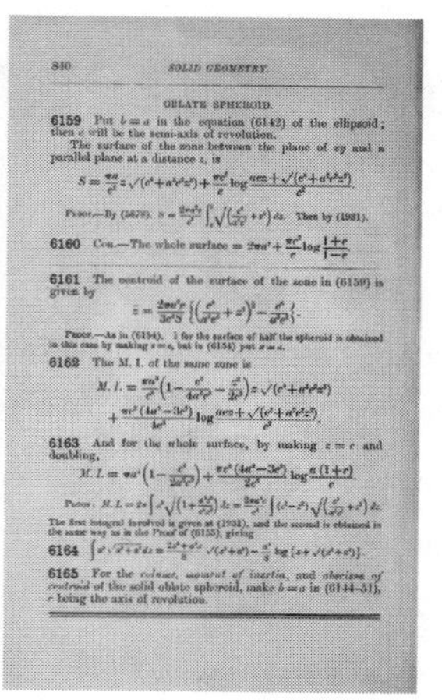

840 SOLID GEOMETRY.

OBLATE SPHEROID.

6159 Put $b = a$ in the equation (6142) of the ellipsoid; then c will be the semi-axis of revolution.

The surface of the zone between the plane of xy and a parallel plane at a distance z, is

6160 COR.—The whole surface $= 2\pi a^2 + \frac{\pi c^2}{e}\log\frac{1+e}{1-e}$

6161 The centroid of the surface of the zone in (6159) is given by

6162 The M. I. of the same zone is

6163 And for the whole surface, by making $z = c$ and doubling,

6165 For the volume, moment of inertia, and abscissa of centroid of the solid oblate spheroid, make $b = a$ in (6144–51), c being the axis of revolution.

By the time Ramanujan got to college, all he wanted to do was mathematics—and he failed his other classes, and at one point ran away, causing his mother to send a missing-person letter to the newspaper:

A MISSING BOY.

To the Editor of the "Hindu."

Sir,—Kindly insert the following in your widely circulated journal:

"A Brahmin boy of the Vaishnava (Thengalai) sect, named Ramanujam, of fair complexion and aged about 18 years was till recently a student of the Kumbakonam College. He left his home on some misunderstanding. His guardian is very solicitous about the boy's returning home. He stayed at Rajahmundry for about a month, and was last seen there some five days back. Those who happen to see him are kindly requested to pursuade him to return home, and to communicate his whereabouts to.

J. Sreenivasa Raghava Ayangar,
18, Sarangapani Sannidhi Street,
Kumbakonam.

September 2.

Ramanujan moved to Madras (now Chennai), tried different colleges, had medical problems, and continued his independent math research. In 1909, when he was 21, his mother arranged (in keeping with customs of the time) for him to marry a then-10-year-old girl named Janaki, who started living with him a couple of years later.

Ramanujan seems to have supported himself by doing math tutoring—but soon became known around Madras as a math whiz, and began publishing in the recently launched *Journal of the Indian Mathematical Society*. His first paper—published in 1911—was on computational properties of Bernoulli numbers (the same Bernoulli numbers that Ada Lovelace had used in her 1843 paper on the Analytical Engine). Though his results weren't spectacular, Ramanujan's approach was an interesting and original one that combined continuous ("what's the numerical value?") and discrete ("what's the prime factorization?") mathematics.

When Ramanujan's mathematical friends didn't succeed in getting him a scholarship, Ramanujan started looking for jobs, and wound up in March 1912 as an accounting clerk—or effectively, a human calculator—for the Port of Madras (which was then, as now, a big shipping hub). His boss, the Chief Accountant, happened to be interested in academic mathematics, and became a lifelong supporter of his. The head of the Port of Madras was a rather distinguished British civil engineer, and partly through him, Ramanujan started interacting with a network of technically oriented British expatriates. They struggled to assess him, wondering whether "he has the stuff of great mathematicians" or whether "his brains are akin to those of the calculating boy". They wrote to a certain Professor M. J. M. Hill in London, who looked at Ramanujan's rather outlandish statements about divergent series and declared that "Mr. Ramanujan is evidently a man with a taste for Mathematics, and with some ability, but he has got on to wrong lines." Hill suggested some books for Ramanujan to study.

Meanwhile, Ramanujan's expat friends were continuing to look for support for him—and he decided to start writing to British mathematicians himself, though with some significant help at composing the English in his letters. We don't know exactly who all he wrote to first—although Hardy's long-time collaborator John Littlewood mentioned two names shortly before he died 64 years later: H. F. Baker and E. W. Hobson. Neither were particularly good choices: Baker worked on algebraic geometry and Hobson on mathematical analysis, both subjects fairly far from what Ramanujan was doing. But in any event, neither of them responded.

And so it was that on Thursday, January 16, 1913, Ramanujan sent his letter to G. H. Hardy.

Who Was Hardy?

G. H. Hardy was born in 1877 to schoolteacher parents based about 30 miles south of London. He was from the beginning a top student, particularly in mathematics. Even when I was growing up in England in the early 1970s, it was typical for such students to go to Winchester for high school and Cambridge for college. And that's exactly what Hardy did. (The other, slightly more famous, track—less austere and less mathematically oriented—was Eton and Oxford, which happens to be where I went.)

Cambridge undergraduate mathematics was at the time very focused on solving ornately constructed calculus-related problems as a kind of competitive sport—with the final event being the Mathematical Tripos exams, which ranked everyone from the "Senior Wrangler" (top score) to the "Wooden Spoon" (lowest passing score). Hardy thought he should have been top, but actually came in 4th, and decided that what he really liked was the somewhat more rigorous and formal approach to mathematics that was then becoming popular in Continental Europe.

The way the British academic system worked at that time—and basically until the 1960s—was that as soon as they graduated, top students could be elected to "college fellowships" that could last the rest of their lives. Hardy was at Trinity College—the largest and most scientifically distinguished college at Cambridge University—and when he graduated in 1900, he was duly elected to a college fellowship.

Hardy's first research paper was about doing integrals like these:

In[1]:= $\int_0^{\frac{\pi}{4}} \text{Log}[\text{Tan}[\theta]]^2 \, d\theta$

Out[1]= $\frac{\pi^3}{16}$

In[2]:= $\int_0^{\infty} \frac{\text{Sinh}[\frac{x}{2}]}{x \, \text{Cosh}[x]} \, dx$

Out[2]= $-\text{Log}\left[\text{Tan}\left[\frac{\pi}{8}\right]\right]$

For a decade Hardy basically worked on the finer points of calculus, figuring out how to do different kinds of integrals and sums, and injecting greater rigor into issues like convergence and the interchange of limits.

His papers weren't grand or visionary, but they were good examples of state-of-the-art mathematical craftsmanship. (As a colleague of Bertrand Russell's, he dipped into the new area of transfinite numbers, but didn't do much with them.) Then in 1908, he wrote a textbook entitled *A Course of Pure Mathematics*—which was a good book, and was very successful in its time, even if its preface began by explaining that it was for students "whose abilities reach or approach something like what is usually described as 'scholarship standard'".

By 1910 or so, Hardy had pretty much settled into a routine of life as a Cambridge professor, pursuing a steady program of academic work. But then he met John Littlewood. Littlewood had grown up in South Africa and was eight years younger than Hardy, a recent Senior Wrangler, and in many ways much more adventurous. And in 1911 Hardy—who had previously always worked on his own—began a collaboration with Littlewood that ultimately lasted the rest of his life.

As a person, Hardy gives me the impression of a good schoolboy who never fully grew up. He seemed to like living in a structured environment, concentrating on his math exercises, and displaying cleverness whenever he could. He could be very nerdy—whether about cricket scores, proving the non-existence of God, or writing down rules for his collaboration with Littlewood. And in a quintessentially British way, he could express himself with wit and charm, but was personally stiff and distant—for example always theming himself as "G. H. Hardy", with "Harold" basically used only by his mother and sister.

So in early 1913 there was Hardy: a respectable and successful, if personally reserved, British mathematician, who had recently been energized by starting to collaborate with Littlewood—and was being pulled in the direction of number theory by Littlewood's interests there. But then he received the letter from Ramanujan.

The Letter and Its Effects

Ramanujan's letter began in a somewhat unpromising way, giving the impression that he thought he was describing for the first time the already fairly well-known technique of analytic continuation for generalizing things like the factorial function to non-integers. He made the statement that "My whole investigations are based upon this and I have been developing this to a remarkable extent so much so that the local mathematicians are not able to understand me in my higher flights." But after the cover letter, there followed more than nine pages that listed over 120 different mathematical results.

Again, they began unpromisingly, with rather vague statements about having a method to count the number of primes up to a given size. But by page 3, there were definite formulas for sums and integrals and things. Some of them looked at least from a distance like the kinds of things that were, for example, in Hardy's papers. But some were definitely more exotic. Their general texture, though, was typical of these types of math formulas. But many of the actual formulas were quite surprising—often claiming that things one wouldn't expect to be related at all were actually mathematically equal.

At least two pages of the original letter have gone missing. But the last page we have again seems to end inauspiciously—with Ramanujan describing achievements of his theory of divergent series, including the seemingly absurd result about adding up all the positive integers, 1+2+3+4+..., and getting −1/12.

So what was Hardy's reaction? First he consulted Littlewood. Was it perhaps a practical joke? Were these formulas all already known, or perhaps completely wrong? Some they recognized, and knew were correct. But many they did not. But as Hardy later said with characteristic clever gloss, they concluded that these too "must be true because, if they were not true, no one would have the imagination to invent them."

Bertrand Russell wrote that by the next day he "found Hardy and Littlewood in a state of wild excitement because they believe they have found a second Newton, a Hindu clerk in Madras making 20 pounds a year." Hardy showed Ramanujan's letter to lots of people, and started making enquiries with the government department that handled India. It took him a week to actually reply to Ramanujan, opening with a certain measured and precisely expressed excitement: "I was exceedingly interested by your letter and by the theorems which you state."

Then he went on: "You will however understand that, before I can judge properly of the value of what you have done, it is essential that I should see proofs of some of your assertions." It was an interesting thing to say. To Hardy, it wasn't enough to know what was true; he wanted to know the proof—the story—of why it was true. Of course, Hardy could have taken it upon himself to find his own proofs. But I think part of it was that he wanted to get an idea of how Ramanujan thought—and what level of mathematician he really was.

His letter went on—with characteristic precision—to group Ramanujan's results into three classes: already known, new and interesting but probably not important, and new and potentially important. But the only things he immediately put in the third category were Ramanujan's statements about counting primes, adding that "almost everything depends on the precise rigour of the methods of proof which you have used."

Hardy had obviously done some background research on Ramanujan by this point, since in his letter he makes reference to Ramanujan's paper on Bernoulli numbers. But in his letter he just says, "I hope very much that you will send me as quickly as possible... a few of your proofs," then closes with, "Hoping to hear from you again as soon as possible."

Ramanujan did indeed respond quickly to Hardy's letter, and his response is fascinating. First, he says he was expecting the same kind of reply from Hardy as he had from the "Mathematics Professor at London", who just told him "not [to] fall into the pitfalls of divergent series." Then he reacts to Hardy's desire for rigorous proofs by saying, "If I had given you my methods of proof I am sure you will follow the London Professor." He mentions his result $1+2+3+4+\ldots=-1/12$ and says that "If I tell you this you will at once point out to me the lunatic asylum as my goal." He goes on to say, "I dilate on this simply to convince you that you will not be able to follow my methods of proof... [based on] a single letter." He says that his first goal is just to get someone like Hardy to verify his results—so he'll be able to get a scholarship, since "I am already a half starving man. To preserve my brains I want food..."

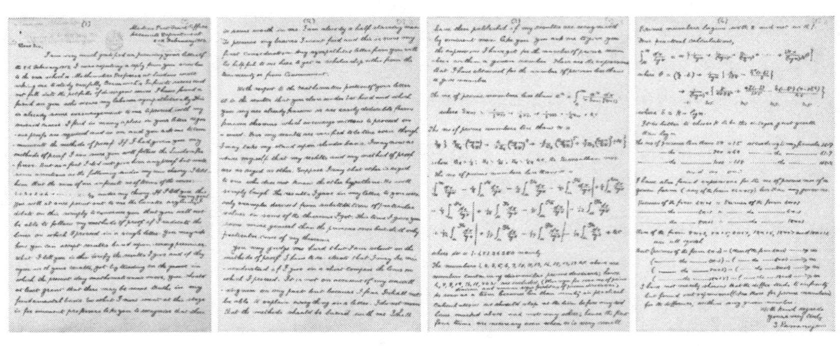

Ramanujan makes a point of saying that it was Hardy's first category of results—ones that were already known—that he's most pleased about, "For my results are verified to be true even though I may take my stand upon slender basis." In other words, Ramanujan himself wasn't sure if the results were correct—and he's excited that they actually are.

So how was he getting his results? I'll say more about this later. But he was certainly doing all sorts of calculations with numbers and formulas—in effect doing experiments. And presumably he was looking at the results of these calculations to get an idea of what might be true. It's not clear how he figured out what was actually true—and indeed some of the results he quoted weren't in the end true. But presumably he used some mixture of traditional mathematical proof, calculational evidence, and lots of intuition. But he didn't explain any of this to Hardy.

Instead, he just started conducting a correspondence about the details of the results, and the fragments of proofs he was able to give. Hardy and Littlewood seemed intent on grading his efforts—with Littlewood writing about some result, for example, "(d) is still wrong, of course, rather a howler." Still, they wondered if Ramanujan was "an Euler", or merely "a Jacobi". But Littlewood had to say, "The stuff about primes is wrong"—explaining that Ramanujan incorrectly assumed the Riemann zeta function didn't have zeros off the real axis, even though it actually has an infinite number of them, which are the subject of the whole Riemann hypothesis. (The Riemann hypothesis is still a famous unsolved math problem, even though an optimistic teacher suggested it to Littlewood as a project when he was an undergraduate...)

What about Ramanujan's strange 1+2+3+4+... = −1/12? Well, that has to do with the Riemann zeta function as well. For positive integers, $\zeta(s)$ is defined as the sum $1/1^s+1/2^s+1/3^s+$.... And given those values, there's a nice function—called Zeta[s] in the Wolfram Language—that can be obtained by continuing to all complex *s*. Now based on the formula for positive arguments, one can identify Zeta[-1] with 1+2+3+4+... But one can also just evaluate Zeta[-1]:

```
In[1]:= Zeta[-1]
```

Out[1]= $-\frac{1}{12}$

It's a weird result, to be sure. But not as crazy as it might at first seem. And in fact it's a result that's nowadays considered perfectly sensible for purposes of certain calculations in quantum field theory (in which, to be fair, all actual infinities are intended to cancel out at the end).

But back to the story. Hardy and Littlewood didn't really have a good mental model for Ramanujan. Littlewood speculated that Ramanujan might not be giving the proofs they assumed he had because he was afraid they'd steal his work. (Stealing was a major issue in academia then as it is now.) Ramanujan said he was "pained" by this speculation, and assured them that he was not "in the least apprehensive of my method being utilised by others." He said that actually he'd invented the method eight years earlier, but hadn't found anyone who could appreciate it, and now he was "willing to place unreservedly in your possession what little I have."

Meanwhile, even before Hardy had responded to Ramanujan's first letter, he'd been investigating with the government department responsible for Indian students how he could bring Ramanujan to Cambridge. It's not quite clear quite what got communicated, but Ramanujan responded that he couldn't go—perhaps because of his Brahmin beliefs, or his mother, or perhaps because he just didn't think he'd fit in. But in any case, Ramanujan's supporters started pushing instead for him to get a graduate scholarship at the University of Madras. More experts were consulted, who opined that "His results appear to be wonderful; but he is not, now, able to present any intelligible proof of some of them," but "He has sufficient knowledge of English and is not too old to learn modern methods from books."

The university administration said their regulations didn't allow a gradate scholarship to be given to someone like Ramanujan who hadn't finished an undergraduate degree. But they helpfully suggested that "Section XV of the Act of Incorporation and Section 3 of the Indian Universities Act, 1904, allow of the grant of such a scholarship [by the Government Educational Department], subject to the express consent of the Governor of Fort St George in Council." And despite the seemingly arcane bureaucracy, things moved quickly, and within a few weeks Ramanujan was duly awarded a scholarship for two years, with the sole requirement that he provide quarterly reports.

A Way of Doing Mathematics

By the time he got his scholarship, Ramanujan had started writing more papers, and publishing them in the *Journal of the Indian Mathematical Society*. Compared to his big claims about primes and divergent series, the topics of these papers were quite tame. But the papers were remarkable nevertheless.

What's immediately striking about them is how calculational they are—full of actual, complicated formulas. Most math papers aren't that way. They may have complicated notation, but they don't have big expressions containing complicated combinations of roots, or seemingly random long integers.

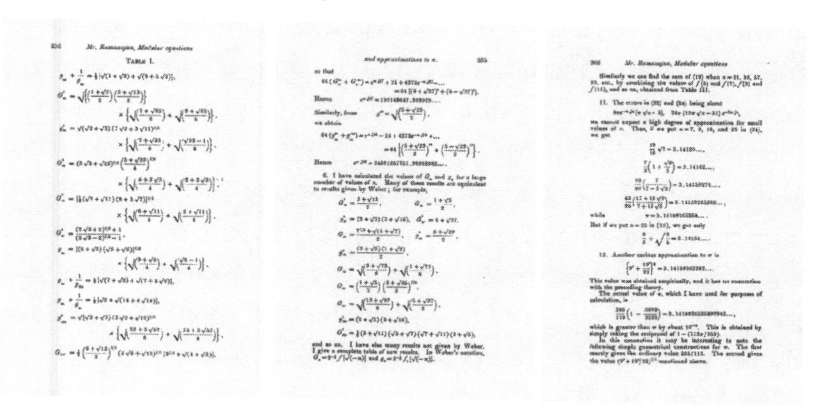

In modern times, we're used to seeing incredibly complicated formulas routinely generated by Mathematica. But usually they're just intermediate steps, and aren't what papers explicitly talk much about. For Ramanujan, though, complicated formulas were often what really told the story. And of course it's incredibly impressive that he could derive them without computers and modern tools.

(As an aside, back in the late 1970s I started writing papers that involved formulas generated by computer. And in one particular paper, the formulas happened to have lots of occurrences of the number 9. But the experienced typist who typed the paper—yes, from a manuscript—replaced every "9" with a "g". When I asked her why, she said, "Well, there are never explicit 9's in papers!")

Looking at Ramanujan's papers, another striking feature is the frequent use of numerical approximations in arguments leading to exact results. People tend to think of working with algebraic formulas as an exact process—generating, for example, coefficients that are exactly 16, not just roughly 15.99999. But for Ramanujan, approximations were routinely part of the story, even when the final results were exact.

In some sense it's not surprising that approximations to numbers are useful. Let's say we want to know which is larger: $\sqrt{2}^{\sqrt{2}+\sqrt{3}}$ or $2^{\sqrt{3}}$. We can start doing all sorts of transformations among square roots, and trying to derive theorems from them. Or we can just evaluate each expression numerically, and find that the first one (2.9755...) is obviously smaller than the second (3.322...). In the mathematical tradition of someone like Hardy—or, for that matter, in a typical modern calculus course—such a direct calculational way of answering the question seems somehow inappropriate and improper.

And of course if the numbers are very close one has to be careful about numerical round-off and so on. But for example in Mathematica and the Wolfram Language today—particularly with their built-in precision tracking for numbers—we often use numerical approximations internally as part of deriving exact results, much like Ramanujan did.

When Hardy asked Ramanujan for proofs, part of what he wanted was to get a kind of story for each result that explained why it was true. But in a sense Ramanujan's methods didn't lend themselves to that. Because part of the "story" would have to be that there's this complicated expression, and it happens to be numerically greater than this other expression. It's easy to see it's true—but there's no real story of why it's true.

And the same happens whenever a key part of a result comes from pure computation of complicated formulas, or in modern times, from automated theorem proving. Yes, one can trace the steps and see that they're correct. But there's no bigger story that gives one any particular understanding of the results.

For most people it'd be bad news to end up with some complicated expression or long seemingly random number—because it wouldn't tell them anything. But Ramanujan was different. Littlewood once said of Ramanujan that "every positive integer was one of his personal friends." And between a good memory and good ability to notice patterns, I suspect Ramanujan could conclude a lot from a complicated expression or a long number. For him, just the object itself would tell a story.

Ramanujan was of course generating all these things by his own calculational efforts. But back in the late 1970s and early 1980s I had the experience of starting to generate lots of complicated results automatically by computer. And after I'd been doing it awhile, something interesting happened: I started being able to quickly recognize the "texture" of results—and often immediately see what might be likely to be true. If I was dealing, say, with some complicated integral, it wasn't that I knew any theorems about it. I just had an intuition about, for example, what functions might appear in the result. And given this, I could then get the computer to go in and fill in the details—and check that the result was correct. But I couldn't derive why the result was true, or tell a story about it; it was just something that intuition and calculation gave me.

Now of course there's a fair amount of pure mathematics where one can't (yet) just routinely go in and do an explicit computation to check whether or not some result is correct. And this often happens for example when there are infinite or infinitesimal quantities or limits involved. And one of the things Hardy had specialized in was giving proofs that were careful in handling such things. In 1910 he'd even written a book called *Orders of Infinity* that was about subtle issues that come up in taking infinite limits. (In particular, in a kind of algebraic analog of the theory of transfinite numbers, he talked about comparing growth rates of things like nested exponential functions—and we even make some use of what are now called Hardy fields in dealing with generalizations of power series in the Wolfram Language.)

So when Hardy saw Ramanujan's "fast and loose" handling of infinite limits and the like, it wasn't surprising that he reacted negatively—and thought he would need to "tame" Ramanujan, and educate him in the finer European ways of doing such things, if Ramanujan was actually going to reliably get correct answers.

Seeing What's Important

Ramanujan was surely a great human calculator, and impressive at knowing whether a particular mathematical fact or relation was actually true. But his greatest skill was, I think, something in a sense more mysterious: an uncanny ability to tell what was significant, and what might be deduced from it.

Take for example his paper "Modular Equations and Approximations to π", published in 1914, in which he calculates (without a computer of course):

$$e^{\pi\sqrt{58}} = 24591257751.99999982\ldots.$$

Most mathematicians would say, "It's an amusing coincidence that that's so close to an integer—but so what?" But Ramanujan realized there was more to it. He found other relations (those "=" should really be "≅"):

$$e^{\frac{1}{8}\pi\sqrt{18}} = 2\sqrt{7}, \quad e^{\pi\sqrt{22}/12} = 2+\sqrt{2}, \quad e^{\frac{1}{4}\pi\sqrt{30}} = 20\sqrt{3}+16\sqrt{6},$$

$$e^{\frac{1}{4}\pi\sqrt{34}} = 12(4+\sqrt{17}), \quad e^{\frac{1}{2}\pi\sqrt{46}} = 144(147+104\sqrt{2}),$$

$$e^{\frac{1}{4}\pi\sqrt{42}} = 84+32\sqrt{6}, \quad e^{\pi\sqrt{58}/12} = \frac{5+\sqrt{29}}{\sqrt{2}},$$

$$e^{\frac{1}{4}\pi\sqrt{70}} = 60\sqrt{35}+96\sqrt{14}, \quad e^{\frac{1}{4}\pi\sqrt{78}} = 300\sqrt{3}+208\sqrt{6},$$

$$e^{\pi\sqrt{55}/24} = \frac{1+\sqrt{(3+2\sqrt{5})}}{\sqrt{2}}, \quad e^{\frac{1}{4}\pi\sqrt{102}} = 800\sqrt{3}+196\sqrt{51},$$

$$e^{\frac{1}{4}\pi\sqrt{130}} = 12(323+40\sqrt{65}), \quad e^{\pi\sqrt{190}/12} = (2\sqrt{2}+\sqrt{10})(3+\sqrt{10}),$$

$$\pi = \frac{12}{\sqrt{130}}\log\left\{\frac{(2+\sqrt{5})(3+\sqrt{13})}{\sqrt{2}}\right\},$$

$$\pi = \frac{24}{\sqrt{142}}\log\left\{\sqrt{\left(\frac{10+11\sqrt{2}}{4}\right)}+\sqrt{\left(\frac{10+7\sqrt{2}}{4}\right)}\right\},$$

$$\pi = \frac{12}{\sqrt{190}}\log\{(2\sqrt{2}+\sqrt{10})(3+\sqrt{10})\}.$$

$$\pi = \frac{12}{\sqrt{310}}\log\left[\tfrac{1}{4}(3+\sqrt{5})(2+\sqrt{2})\{(5+2\sqrt{10})+\sqrt{(61+20\sqrt{10})}\}\right].$$

$$\pi = \frac{4}{\sqrt{522}}\log\left[\left(\frac{5+\sqrt{29}}{\sqrt{2}}\right)^3(5\sqrt{29}+11\sqrt{6}) \times \left\{\sqrt{\left(\frac{9+3\sqrt{6}}{4}\right)}+\sqrt{\left(\frac{5+3\sqrt{6}}{4}\right)}\right\}^6\right].$$

Then he began to build a theory—that involves elliptic functions, though Ramanujan didn't know that name yet—and started coming up with new series approximations for π:

$$(28)\quad \frac{4}{\pi}=1+\frac{7}{4}\left(\frac{1}{2}\right)^3+\frac{13}{4^2}\left(\frac{1.3}{2.4}\right)^3+\frac{19}{4^3}\left(\frac{1.3.5}{2.4.6}\right)^3+\ldots,$$
$$(q=e^{-\pi\sqrt{3}},\ 2kk'=\tfrac{1}{2}),$$

$$(29)\quad \frac{16}{\pi}=5+\frac{47}{64}\left(\frac{1}{2}\right)^3+\frac{89}{64^2}\left(\frac{1.3}{2.4}\right)^3+\frac{131}{64^3}\left(\frac{1.3.5}{2.4.6}\right)^3+\ldots,$$
$$(q=e^{-\pi\sqrt{7}},\ 2kk'=\tfrac{1}{8}),$$

$$(30)\quad \frac{32}{\pi}=(5\sqrt{5}-1)+\frac{47\sqrt{5}+29}{64}\left(\frac{1}{2}\right)^3\left(\frac{\sqrt{5}-1}{2}\right)^8+\ldots,$$
$$\frac{89\sqrt{5}+59}{64^2}\left(\frac{1.3}{2.4}\right)^3\left(\frac{\sqrt{5}-1}{2}\right)^{16}+\ldots,$$
$$\left[q=e^{-\pi\sqrt{15}},\ 2kk'=\frac{1}{8}\left(\frac{\sqrt{5}-1}{2}\right)\right];$$

$$(42)\quad \frac{1}{3\pi\sqrt{3}}=\frac{3}{49}+\frac{43}{49^3}\cdot\frac{1}{2}\cdot\frac{1.3}{4^2}+\frac{83}{49^5}\cdot\frac{1.3}{2.4}\cdot\frac{1.3.5.7}{4^2.8^2}+\ldots,$$

$$(43)\quad \frac{2}{\pi\sqrt{11}}=\frac{19}{99}+\frac{299}{99^3}\cdot\frac{1}{2}\cdot\frac{1.3}{4^2}+\frac{579}{99^5}\cdot\frac{1.3}{2.4}\cdot\frac{1.3.5.7}{4^2.8^2}+\ldots,$$

$$(44)\quad \frac{1}{2\pi\sqrt{2}}=\frac{1103}{99^2}+\frac{27493}{99^6}\cdot\frac{1}{2}\cdot\frac{1.3}{4^2}+\frac{53883}{99^{10}}\cdot\frac{1.3}{2.4}\cdot\frac{1.3.5.7}{4^2.8^2}+\ldots.$$

Previous approximations to π had in a sense been much more sober, though the best one before Ramanujan's (Machin's series from 1706) did involve the seemingly random number 239:

$$\pi=16\left(\frac{1}{5}-\frac{1}{3\times5^3}+\frac{1}{5\times5^5}-\ldots\right)-4\left(\frac{1}{239}-\frac{1}{3\times239^3}+\frac{1}{5\times239^5}-\ldots\right)$$

But Ramanujan's series—bizarre and arbitrary as they might appear—had an important feature: they took far fewer terms to compute π to a given accuracy. In 1977, Bill Gosper—himself a rather Ramanujan-like figure, whom I've had the pleasure of knowing for more than 35 years—took the last of Ramanujan's series from the list above, and used it to compute a record number of digits of π. There soon followed other computations, all based directly on Ramanujan's idea—as is the method we use for computing π in Mathematica and the Wolfram Language.

It's interesting to see in Ramanujan's paper that even he occasionally didn't know what was and wasn't significant. For example, he noted:

12. Another curious approximation to π is

$$\left\{9^2 + \frac{19^2}{22}\right\}^{\frac{1}{4}} = 3.14159265262\ldots.$$

This value was obtained empirically, and it has no connection with the preceding theory.

And then—in pretty much his only published example of geometry—he gave a peculiar geometric construction for approximately "squaring the circle" based on this formula:

Then the square on BX is very nearly equal to the area of the circle, the error being less than a tenth of an inch when the diameter is 40 miles long.

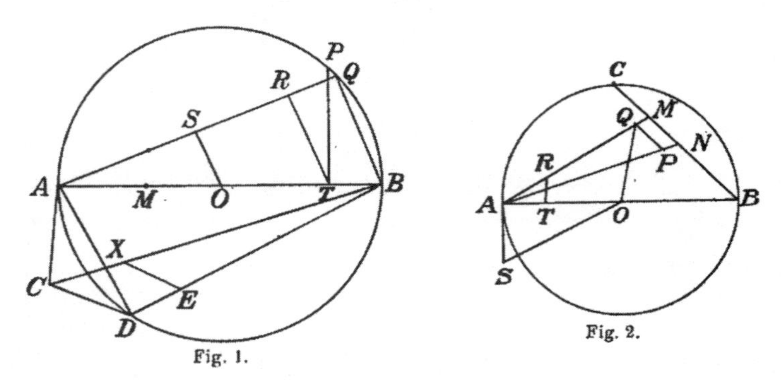

Fig. 1.

Fig. 2.

Truth versus Narrative

To Hardy, Ramanujan's way of working must have seemed quite alien. For Ramanujan was in some fundamental sense an experimental mathematician: going out into the universe of mathematical possibilities and doing calculations to find interesting and significant facts—and only then building theories based on them.

Hardy on the other hand worked like a traditional mathematician, progressively extending the narrative of existing mathematics. Most of his papers begin—explicitly or implicitly—by quoting some result from the mathematical literature, and then proceed by telling the story of how this result can be extended by a series of rigorous steps. There are no sudden empirical discoveries—and no seemingly inexplicable jumps based on intuition from them. It's mathematics carefully argued, and built, in a sense, brick by brick.

A century later this is still the way almost all pure mathematics is done. And even if it's discussing the same subject matter, perhaps anything else shouldn't be called "mathematics", because its methods are too different. In my own efforts to explore the computational universe of simple programs, I've certainly done a fair amount that could be called "mathematical" in the sense that it, for example, explores systems based on numbers.

Over the years, I've found all sorts of results that seem interesting. Strange structures that arise when one successively adds numbers to their digit reversals. Bizarre nested recurrence relations that generate primes. Peculiar representations of integers using trees of bitwise XORs. But they're empirical facts—demonstrably true, yet not part of the tradition and narrative of existing mathematics.

For many mathematicians—like Hardy—the process of proof is the core of mathematical activity. It's not particularly significant to come up with a conjecture about what's true; what's significant is to create a proof that explains why something is true, constructing a narrative that other mathematicians can understand.

Particularly today, as we start to be able to automate more and more proofs, they can seem a bit like mundane manual labor, where the outcome may be interesting but the process of getting there is not. But proofs can also be illuminating. They can in effect be stories that introduce new abstract concepts that transcend the particulars of a given proof, and provide raw material to understand many other mathematical results.

For Ramanujan, though, I suspect it was facts and results that were the center of his mathematical thinking, and proofs felt a bit like some strange European custom necessary to take his results out of his particular context, and convince European mathematicians that they were correct.

Going to Cambridge

But let's return to the story of Ramanujan and Hardy.

In the early part of 1913, Hardy and Ramanujan continued to exchange letters. Ramanujan described results; Hardy critiqued what Ramanujan said, and pushed for proofs and traditional mathematical presentation. Then there was a long gap, but finally in December 1913, Hardy wrote again, explaining that Ramanujan's most ambitious results—about the distribution of primes—were definitely incorrect, commenting that "...the theory of primes is full of pitfalls, to surmount which requires the fullest of trainings in modern rigorous methods." He also said that if Ramanujan had been able to prove his results it would have been "about the most remarkable mathematical feat in the whole history of mathematics."

In January 1914 a young Cambridge mathematician named E. H. Neville came to give lectures in Madras, and relayed the message that Hardy was (in Ramanujan's words) "anxious to get [Ramanujan] to Cambridge". Ramanujan responded that back in February 1913 he'd had a meeting, along with his "superior officer", with the Secretary to the Students Advisory Committee of Madras, who had asked whether he was prepared to go to England. Ramanujan wrote that he assumed he'd have to take exams like the other Indian students he'd seen go to England, which he didn't think he'd do well enough in—and also that his superior officer, a "very orthodox Brahman having scruples to go to foreign land replied at once that I could not go".

But then he said that Neville had "cleared [his] doubts", explaining that there wouldn't be an issue with his expenses, that his English would do, that he wouldn't have to take exams, and that he could remain a vegetarian in England. He ended by saying that he hoped Hardy and Littlewood would "be good enough to take the trouble of getting me [to England] within a very few months."

Hardy had assumed it would be bureaucratically trivial to get Ramanujan to England, but actually it wasn't. Hardy's own Trinity College wasn't prepared to contribute any real funding. Hardy and Littlewood offered to put up some of the money themselves. But Neville wrote to the registrar of the University of Madras saying that "the discovery of the genius of S. Ramanujan of Madras promises to be the most interesting event of our time in the mathematical world"—and suggested the university come up with the money. Ramanujan's expat supporters swung into action, with the matter eventually reaching the Governor of Madras—and a solution was found that involved taking money from a grant that had been given by the government five years earlier for "establishing University vacation lectures", but that was actually, in the bureaucratic language of "Document No. 182 of the Educational Department", "not being utilised for any immediate purpose".

There are strange little notes in the bureaucratic record, like on February 12: "What caste is he? Treat as urgent." But eventually everything was sorted out, and on March 17, 1914, after a send-off featuring local dignitaries, Ramanujan boarded a ship for England, sailing up through the Suez Canal, and arriving in London on April 14. Before leaving India, Ramanujan had prepared for European life by getting Western clothes, and learning things like how to eat with a knife and fork, and how to tie a tie. Many Indian students had come to England before, and there was a whole procedure for them. But after a few days in London, Ramanujan arrived in Cambridge—with the Indian newspapers proudly reporting that "Mr. S. Ramanujan, of Madras, whose work in the higher mathematics has excited the wonder of Cambridge, is now in residence at Trinity."

(In addition to Hardy and Littlewood, two other names that appear in connection with Ramanujan's early days in Cambridge are Neville and Barnes. They're not especially famous in the overall history of mathematics, but it so happens that in the Wolfram Language they're both commemorated by built-in functions: NevilleThetaS and BarnesG.)

Ramanujan in Cambridge

What was the Ramanujan who arrived in Cambridge like? He was described as enthusiastic and eager, though diffident. He made jokes, sometimes at his own expense. He could talk about politics and philosophy as well as mathematics. He was never particularly introspective. In official settings he was polite and deferential and tried to follow local customs. His native language was Tamil, and earlier in his life he had failed English exams, but by the time he arrived in England, his English was excellent. He liked to hang out with other Indian students, sometimes going to musical events, or boating on the river. Physically, he was described as short and stout—with his main notable feature being the brightness of his eyes. He worked hard, chasing one mathematical problem after another. He kept his living space sparse, with only a few books and papers. He was sensible about practical things, for example in figuring out issues with cooking and vegetarian ingredients. And from what one can tell, he was happy to be in Cambridge.

But then on June 28, 1914—two and a half months after Ramanujan arrived in England—Archduke Ferdinand was assassinated, and on July 28, World War I began. There was an immediate effect on Cambridge. Many students were called up for military duty. Littlewood joined the war effort and ended up developing ways to compute range tables for anti-aircraft guns. Hardy wasn't a big supporter of the war—not least because he liked German mathematics—but he volunteered for duty too, though was rejected on medical grounds.

Ramanujan described the war in a letter to his mother, saying for example, "They fly in aeroplanes at great heights, bomb the cities and ruin them. As soon as enemy planes are sighted in the sky, the planes resting on the ground take off and fly at great speeds and dash against them resulting in destruction and death."

Ramanujan nevertheless continued to pursue mathematics, explaining to his mother that "war is waged in a country that is as far as Rangoon is away from [Madras]". There were practical difficulties, like a lack of vegetables, which caused Ramanujan to ask a friend in India to send him "some tamarind (seeds being removed) and good cocoanut oil by parcel post". But of more importance, as Ramanujan reported it, was that the "professors here... have lost their interest in mathematics owing to the present war".

Ramanujan told a friend that he had "changed [his] plan of publishing [his] results". He said that he would wait to publish any of the old results in his notebooks until the war was over. But he said that since coming to England he had learned "their methods", and was "trying to get new results by their methods so that I can easily publish these results without delay".

In 1915 Ramanujan published a long paper entitled "Highly Composite Numbers" about maxima of the function (DivisorSigma in the Wolfram

Language) that counts the number of divisors of a given number. Hardy seems to have been quite involved in the preparation of this paper—and it served as the centerpiece of Ramanujan's analog of a PhD thesis.

For the next couple of years, Ramanujan prolifically wrote papers—and despite the war, they were published. A notable paper he wrote with Hardy concerns the partition function (PartitionsP in the Wolfram Language) that counts the number of ways an integer can be written as a sum of positive integers. The paper is a classic example of mixing the approximate with the exact. The paper begins with the result for large *n*:

$$p(n) \sim P(n) = \frac{1}{4\pi\sqrt{3}} e^{\pi\sqrt{\frac{2n}{3}}}$$

But then, using ideas Ramanujan developed back in India, it progressively improves the estimate, to the point where the exact integer result can be obtained. In Ramanujan's day, computing the exact value of PartitionsP[200] was a big deal—and the climax of his paper. But now, thanks to Ramanujan's method, it's instantaneous:

```
In[1]:= PartitionsP[200]
Out[1]= 3972999029388
```

Cambridge was dispirited by the war—with an appalling number of its finest students dying, often within weeks, at the front lines. Trinity College's big quad had become a war hospital. But through all of this, Ramanujan continued to do his mathematics—and with Hardy's help continued to build his reputation.

But then in May 1917, there was another problem: Ramanujan got sick. From what we know now, it's likely that what he had was a parasitic liver infection picked up in India. But back then nobody could diagnose it. Ramanujan went from doctor to doctor, and nursing home to nursing home. He didn't believe much of what he was told, and nothing that was done seemed to help much. Some months he would be well enough to do a significant amount of mathematics; others not. He became depressed, and at one point apparently suicidal. It didn't help that his mother had prevented his wife back in India from communicating with him, presumably fearing it would distract him.

Hardy tried to help—sometimes by interacting with doctors, sometimes by providing mathematical input. One doctor told Hardy he suspected "some obscure Oriental germ trouble imperfectly studied at present". Hardy wrote, "Like all Indians, [Ramanujan] is fatalistic, and it is terribly hard to get him to take care of himself." Hardy later told the now-famous story that he once visited Ramanujan at a nursing home, telling him that he came in a taxicab with number 1729, and saying that it seemed to him a rather dull number—to which Ramanujan replied: "No, it is a very interesting number; it is the smallest number expressible as the sum of two cubes in two different ways": $1729=1^3+12^3=9^3+10^3$. (Wolfram|Alpha now reports some other properties too.)

But through all of this, Ramanujan's mathematical reputation continued to grow. He was elected a Fellow of the Royal Society (with his supporters including Hobson and Baker, both of whom had failed to respond to his original letter)—and in October 1918 he was elected a fellow of Trinity College, assuring him financial support. A month later World War I was over—and the threat of U-boat attacks, which had made travel to India dangerous, was gone.

And so on March 13, 1919, Ramanujan returned to India—now very famous and respected, but also very ill. Through it all, he continued to do mathematics, writing a notable letter to Hardy about "mock" theta functions on January 12, 1920. He chose to live humbly, and largely ignored what little medicine could do for him. And on April 26, 1920, at the age of 32, and three days after the last entry in his notebook, he died.

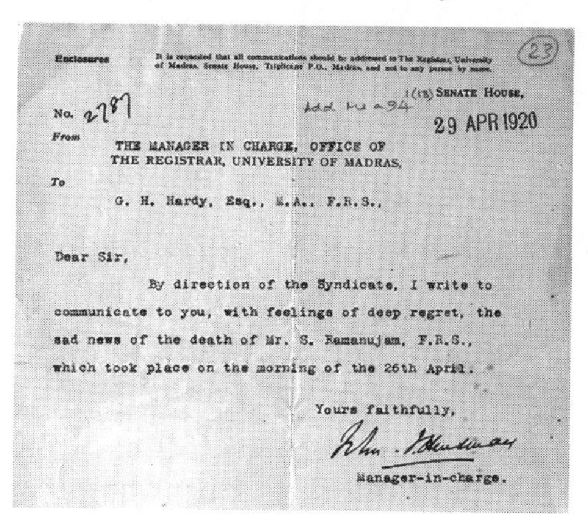

(23)

Enclosures — It is requested that all communications should be addressed to The Registrar, University of Madras, Senate House, Triplicane P.O., Madras, and not to any person by name.

No. 2787 — Add M a 94

SENATE HOUSE,

29 APR 1920

From THE MANAGER IN CHARGE, OFFICE OF THE REGISTRAR, UNIVERSITY OF MADRAS,

To G. H. Hardy, Esq., M.A., F.R.S.,

Dear Sir,

By direction of the Syndicate, I write to communicate to you, with feelings of deep regret, the sad news of the death of Mr. S. Ramanujam, F.R.S., which took place on the morning of the 26th April.

Yours faithfully,

[signature]

Manager-in-charge.

The Aftermath

From when he first started doing mathematics research, Ramanujan had recorded his results in a series of hardcover notebooks—publishing only a very small fraction of them. When Ramanujan died, Hardy began to organize an effort to study and publish all 3000 or so results in Ramanujan's notebooks. Several people were involved in the 1920s and 1930s, and quite a few publications were generated. But through various misadventures the project was not completed—to be taken up again only in the 1970s.

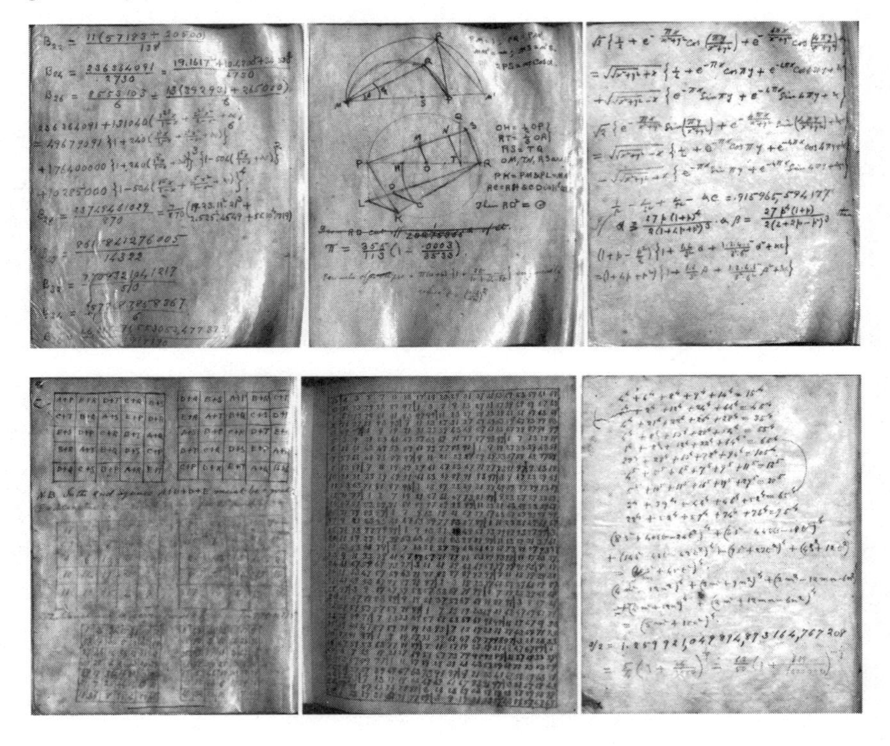

In 1940, Hardy gave all the letters he had from Ramanujan to the Cambridge University Library, but the original cover letter for what Ramanujan sent in 1913 was not among them—so now the only record we have of that is the transcription Hardy later published. Ramanujan's three main notebooks sat for many years on top of a cabinet in the librarian's office at the University of Madras, where they suffered damage from insects, but were never lost. His other mathematical documents passed through several hands, and some of them wound up in the incredibly messy office of a Cambridge mathematician—but when he died in 1965 they were noticed and sent to a library, where they languished until they were "rediscovered" with great excitement as Ramanujan's lost notebook in 1976.

When Ramanujan died, it took only days for his various relatives to start asking for financial support. There were large medical bills from England, and there was talk of selling Ramanujan's papers to raise money.

Ramanujan's wife was 21 when he died, but as was the custom, she never remarried. She lived very modestly, making her living mostly from tailoring. In 1950 she adopted the son of a friend of hers who had died. By the 1960s, Ramanujan was becoming something of a general Indian hero, and she started receiving various honors and pensions. Over the years, quite a few mathematicians had come to visit her—and she had supplied them for example with the passport photo that has become the most famous picture of Ramanujan.

She lived a long life, dying in 1994 at the age of 95, having outlived Ramanujan by 73 years.

What Became of Hardy?

Hardy was 35 when Ramanujan's letter arrived, and was 43 when Ramanujan died. Hardy viewed his "discovery" of Ramanujan as his greatest achievement, and described his association with Ramanujan as the "one romantic incident of [his] life". After Ramanujan died, Hardy put some of his efforts into continuing to decode and develop Ramanujan's results, but for the most part he returned to his previous mathematical trajectory. His collected works fill seven large volumes (while Ramanujan's publications make up just one fairly slim volume). The word clouds of the titles of his papers show only a few changes from before he met Ramanujan to after:

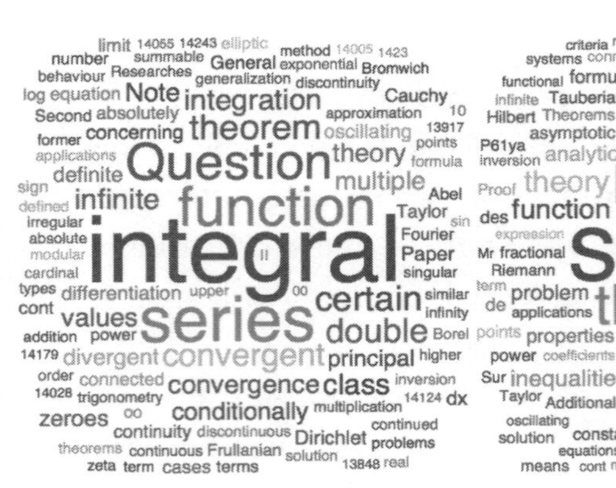

Shortly before Ramanujan entered his life, Hardy had started to collaborate with John Littlewood, who he would later say was an even more important influence on his life than Ramanujan. After Ramanujan died, Hardy moved to what seemed like a better job in Oxford, and ended up staying there for 11 years before returning to Cambridge. His absence didn't affect his collaboration with Littlewood, though—since they worked mostly by exchanging written messages, even when their rooms were less than a hundred feet apart. After 1911 Hardy rarely did mathematics without a collaborator; he worked especially with Littlewood, publishing 95 papers with him over the course of 38 years.

Hardy's mathematics was always of the finest quality. He dreamed of doing something like solving the Riemann hypothesis—but in reality never did anything truly spectacular. He wrote two books, though, that continue to be read today: *An Introduction to the Theory of Numbers*, with E. M. Wright; and *Inequalities*, with Littlewood and G. Pólya.

Hardy lived his life in the stratum of the intellectual elite. In the 1920s he displayed a picture of Lenin in his apartment, and was briefly president of the "scientific workers" trade union. He always wrote elegantly, mostly about mathematics, and sometimes about Ramanujan. He eschewed gadgets and always lived along with students and other professors in his college. He never married, though near the end of his life his younger sister joined him in Cambridge (she also had never married, and had spent most of her life teaching at the girls' school where she went as a child).

In 1940 Hardy wrote a small book called *A Mathematician's Apology*. I remember when I was about 12 being given a copy of this book. I think many people viewed it as a kind of manifesto or advertisement for pure mathematics. But I must say it didn't resonate with me at all. It felt to me at once sanctimonious and austere, and I wasn't impressed by its attempt to describe the aesthetics and pleasures of mathematics, or by the pride with which its author said that "nothing I have ever done is of the slightest practical use" (actually, he co-invented the Hardy–Weinberg law used in genetics). I doubt I would have chosen the path of a pure mathematician anyway, but Hardy's book helped make certain of it.

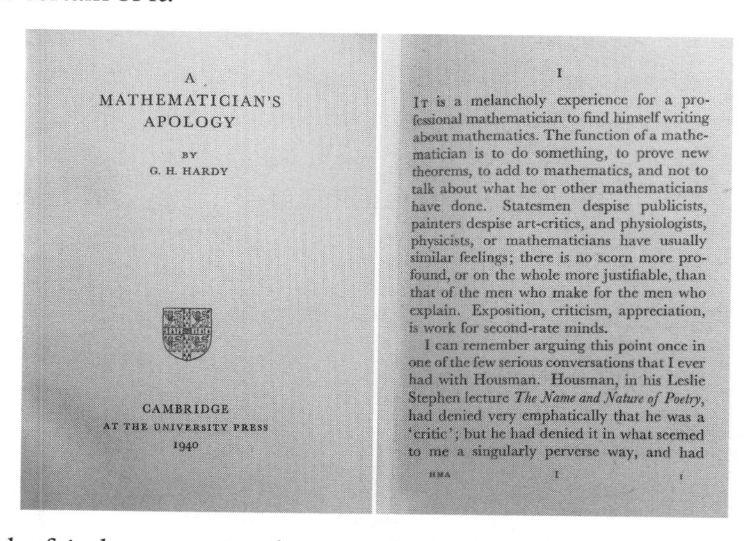
A
MATHEMATICIAN'S
APOLOGY

BY
G. H. HARDY

CAMBRIDGE
AT THE UNIVERSITY PRESS
1940

I

It is a melancholy experience for a professional mathematician to find himself writing about mathematics. The function of a mathematician is to do something, to prove new theorems, to add to mathematics, and not to talk about what he or other mathematicians have done. Statesmen despise publicists, painters despise art-critics, and physiologists, physicists, or mathematicians have usually similar feelings; there is no scorn more profound, or on the whole more justifiable, than that of the men who make for the men who explain. Exposition, criticism, appreciation, is work for second-rate minds.

I can remember arguing this point once in one of the few serious conversations that I ever had with Housman. Housman, in his Leslie Stephen lecture *The Name and Nature of Poetry*, had denied very emphatically that he was a 'critic'; but he had denied it in what seemed to me a singularly perverse way, and had

HMA 1 1

To be fair, however, Hardy wrote the book at a low point in his own life, when he was concerned about his health and the loss of his mathematical faculties. And perhaps that explains why he made a point of explaining that "mathematics... is a young man's game". (And in an article about Ramanujan, he wrote that "a mathematician is often comparatively old at 30, and his death may be less of a catastrophe than it seems.") I don't know if the sentiment had been expressed before—but by the 1970s it was taken as an established fact, extending to science as well as mathematics. Kids I knew would tell me I'd better get on with things, because it'd be all over by age 30.

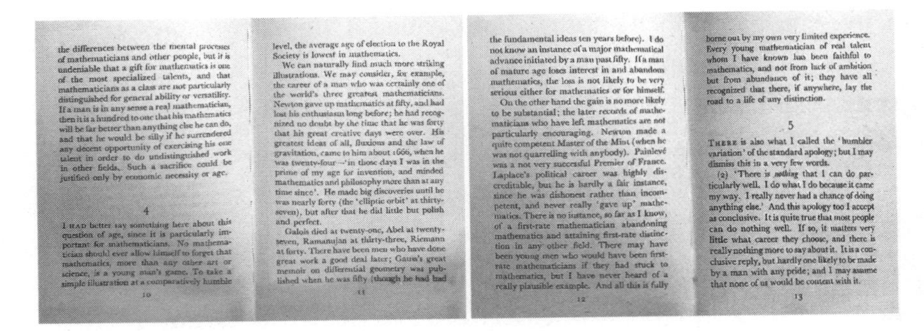

the differences between the mental processes of mathematicians and other people, but it is undeniable that a gift for mathematics is one of the most specialized talents, and that mathematicians as a class are not particularly distinguished for general ability or versatility. If a man is in any sense a real mathematician, then it is a hundred to one that his mathematics will be far better than anything else he can do, and that he would be silly if he surrendered any decent opportunity of exercising his one talent in order to do undistinguished work in other fields. Such a sacrifice could be justified only by economic necessity or age.

4

I HAD better say something here about this question of age, since it is particularly important for mathematicians. No mathematician should ever allow himself to forget that mathematics, more than any other art or science, is a young man's game. To take a simple illustration at a comparatively humble

10

level, the average age of election to the Royal Society is lowest in mathematics.

We can naturally find much more striking illustrations. We may consider, for example, the career of a man who was certainly one of the world's three greatest mathematicians. Newton gave up mathematics at fifty, and had lost his enthusiasm long before; he had recognized no doubt by the time that he was forty that his great creative days were over. His greatest ideas of all, fluxions and the law of gravitation, came to him about 1666, when he was twenty-four—'in those days I was in the prime of my age for invention, and minded mathematics and philosophy more than at any time since'. He made big discoveries until he was nearly forty (the 'elliptic orbit' at thirty-seven), but after that he did little but polish and perfect.

Galois died at twenty-one, Abel at twenty-seven, Ramanujan at thirty-three, Riemann at forty. There have been men who have done great work a good deal later; Gauss's great memoir on differential geometry was published when he was fifty (though he had had

11

the fundamental ideas ten years before). I do not know an instance of a major mathematical advance initiated by a man past fifty. If a man of mature age loses interest in and abandons mathematics, the loss is not likely to be very serious either for mathematics or for himself.

On the other hand the gain is no more likely to be substantial; the later records of mathematicians who have left mathematics are not particularly encouraging. Newton made a quite competent Master of the Mint (when he was not quarrelling with anybody). Painlevé was a not very successful Premier of France. Laplace's political career was highly discreditable, but he is hardly a fair instance, since he was dishonest rather than incompetent, and never really 'gave up' mathematics. There is no instance, so far as I know, of a first-rate mathematician abandoning mathematics and attaining first-rate distinction in any other field. There may have been young men who would have been first-rate mathematicians if they had stuck to mathematics, but I have never heard of a really plausible example. And all this is fully

12

borne out by my own very limited experience. Every young mathematician of real talent whom I have known has been faithful to mathematics, and not from lack of ambition but from abundance of it; they have all recognized that there, if anywhere, lay the road to a life of any distinction.

5

THERE is also what I called the 'humbler variation' of the standard apology; but I may dismiss this in a very few words.

(2) 'There is *nothing* that I can do particularly well. I do what I do because it came my way. I really never had a chance of doing anything else.' And this apology too I accept as conclusive. It is quite true that most people can do nothing well. If so, it matters very little what career they choose, and there is really nothing more to say about it. It is a conclusive reply, but hardly one likely to be made by a man with any pride; and I may assume that none of us would be content with it.

13

Is that actually true? I don't think so. It's hard to get clear evidence, but as one example I took the data we have on notable mathematical theorems in Wolfram|Alpha and the Wolfram Language, and made a histogram of the ages of people who proved them. It's not a completely uniform distribution (though the peak just before 40 is probably just a theorem-selection effect associated with Fields Medals), but particularly if one corrects for life expectancies now and in the past it's a far cry from showing that mathematical productivity has all but dried up by age 30.

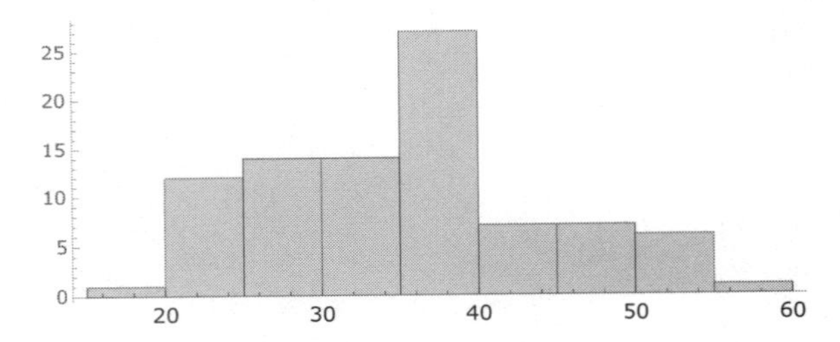

My own feeling—as someone who's getting older myself—is that at least up to my age, many aspects of scientific and technical productivity actually steadily increase. For a start, it really helps to know more—and certainly a lot of my best ideas have come from making connections between things I've learned decades apart. It also helps to have more experience and intuition about how things will work out. And if one has earlier successes, those can help provide the confidence to move forward more definitively, without second guessing. Of course, one must maintain the constitution to focus with enough intensity—and be able to concentrate for long enough—to think through complex things. I think in some ways I've gotten slower over the years, and in some ways faster. I'm slower because I know more about mistakes I make, and try to do things carefully enough to avoid them. But I'm faster because I know more and can shortcut many more things. Of course, for me in particular, it also helps that over the years I've built all sorts of automation that I've been able to make use of.

A quite different point is that while making specific contributions to an existing area (as Hardy did) is something that can potentially be done by the young, creating a whole new structure tends to require the broader knowledge and experience that comes with age.

But back to Hardy. I suspect it was a lack of motivation rather than ability, but in his last years, he became quite dispirited and all but dropped mathematics. He died in 1947 at the age of 70.

Littlewood, who was a decade younger than Hardy, lived on until 1977. Littlewood was always a little more adventurous than Hardy, a little less austere, and a little less august. Like Hardy, he never married—though he did have a daughter (with the wife of the couple who shared his vacation home) whom he described as his "niece" until she was in her forties. And—giving a lie to Hardy's claim about math being a young man's game—Littlewood (helped by getting early antidepressant drugs at the age of 72) had remarkably productive years of mathematics in his 80s.

Ramanujan's Mathematics

What became of Ramanujan's mathematics? For many years, not too much. Hardy pursued it some, but the whole field of number theory—which was where the majority of Ramanujan's work was concentrated—was out of fashion. Here's a plot of the fraction of all math papers tagged as "number theory" as a function of time in the Zentralblatt database:

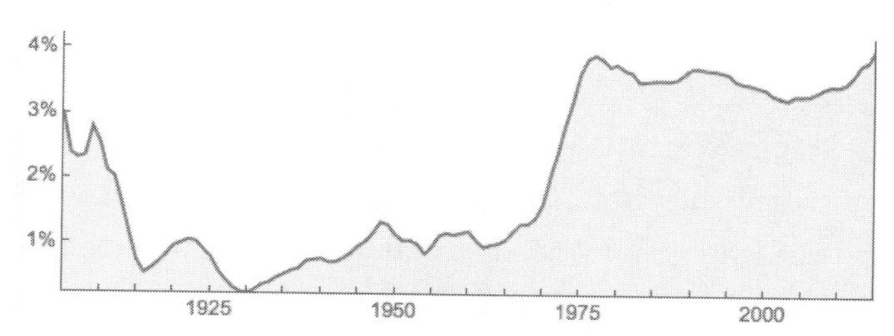

Ramanujan's interest may have been to some extent driven by the peak in the early 1900s (which would probably go even higher with earlier data). But by the 1930s, the emphasis of mathematics had shifted away from what seemed like particular results in areas like number theory and calculus, towards the greater generality and formality that seemed to exist in more algebraic areas.

In the 1970s, though, number theory suddenly became more popular again, driven by advances in algebraic number theory. (Other subcategories showing substantial increases at that time include automorphic forms, elementary number theory and sequences.)

Back in the late 1970s, I had certainly heard of Ramanujan—though more in the context of his story than his mathematics. And I was pleased in 1982, when I was writing about the vacuum in quantum field theory, that I could use results of Ramanujan's to give closed forms for particular cases (of infinite sums in various dimensions of modes of a quantum field—corresponding to Epstein zeta functions):

Some special values of $Z_p(s) \equiv Z_p(a_1 = 1, \ldots, a_p = 1; s)$ are [22]:

$$Z_0(s) = 0,$$

$$Z_1(s) = 2\zeta(s),$$

$$Z_2(s) = 4\zeta\left(\frac{s}{2}\right)\beta\left(\frac{s}{2}\right),$$

$$Z_4(s) = 8(1 - 2^{2-s})\zeta\left(\frac{s}{2}\right)\zeta\left(\frac{s}{2} - 1\right),$$

$$Z_8(s) = 16(1 - 2^{1-s/2} + 2^{4-s})\zeta\left(\frac{s}{2}\right)\zeta\left(\frac{s}{2} - 3\right), \tag{3.5}$$

where $\beta(s) = \sum_{n=0}^{\infty} (-1)^n (2n+1)^{-s}$ and $\beta(1) = \pi/4$, $\beta(2) = G \simeq 0.915$ (G is Catalan's constant), $\beta(3) = \pi^3/32$. $Z_3(s)$ apparently cannot be expressed as a product of one-dimensional sums.

Starting in the 1970s, there was a big effort—still not entirely complete—to prove results Ramanujan had given in his notebooks. And there were increasing connections being found between the particular results he'd got, and general themes emerging in number theory.

A significant part of what Ramanujan did was to study so-called special functions—and to invent some new ones. Special functions—

like the zeta function, elliptic functions, theta functions, and so on—can be thought of as defining convenient "packets" of mathematics. There are an infinite number of possible functions one can define, but what get called "special functions" are ones whose definitions survive because they turn out to be repeatedly useful.

And today, for example, in Mathematica and the Wolfram Language we have RamanujanTau, RamanujanTauL, RamanujanTauTheta and RamanujanTauZ as special functions. I don't doubt that in the future we'll have more Ramanujan-inspired functions. In the last year of his life, Ramanujan defined some particularly ambitious special functions that he called "mock theta functions"—and that are still in the process of being made concrete enough to routinely compute.

If one looks at the definition of Ramanujan's tau function it seems quite bizarre (notice the "24"):

$$\sum_{1}^{\infty} \tau(n)\, x^n = x\{(1-x)(1-x^2)(1-x^3)\ldots\}^{24}$$

And to my mind, the most remarkable thing about Ramanujan is that he could define something as seemingly arbitrary as this, and have it turn out to be useful a century later.

Are They Random Facts?

In antiquity, the Pythagoreans made much of the fact that 1+2+3+4=10. But to us today, this just seems like a random fact of mathematics, not of any particular significance. When I look at Ramanujan's results, many of them also seem like random facts of mathematics. But the amazing thing that's emerged over the past century, and particularly over the past few decades, is that they're not. Instead, more and more of them are being found to be connected to deep, elegant mathematical principles.

To enunciate these principles in a direct and formal way requires layers of abstract mathematical concepts and language which have taken decades to develop. But somehow, through his experiments and intuition, Ramanujan managed to find concrete examples of these principles. Often his examples look quite arbitrary—full of seemingly random definitions and numbers. But perhaps it's not surprising that that's what it takes to express modern abstract principles in terms of the concrete mathematical constructs of the early twentieth century. It's a bit like a poet trying to express deep general ideas—but being forced to use only the imperfect medium of human natural language.

It's turned out to be very challenging to prove many of Ramanujan's results. And part of the reason seems to be that to do so—and to create the kind of narrative needed for a good proof—one actually has no choice but to build up much more abstract and conceptually complex structures, often in many steps.

So how is it that Ramanujan managed in effect to predict all these deep principles of later mathematics? I think there are two basic logical possibilities. The first is that if one drills down from any sufficiently surprising result, say in number theory, one will eventually reach a deep principle in the effort to explain it. And the second possibility is that while Ramanujan did not have the wherewithal to express it directly, he had what amounts to an aesthetic sense of which seemingly random facts would turn out to fit together and have deeper significance.

I'm not sure which of these possibilities is correct, and perhaps it's a combination. But to understand this a little more, we should talk about the overall structure of mathematics. In a sense mathematics as it's practiced is strangely perched between the trivial and the impossible. At an underlying level, mathematics is based on simple axioms. And it could be—as it is, say, for the specific case of Boolean algebra—that given the axioms there's a straightforward procedure to figure out whether any particular result is true. But ever since Gödel's theorem in 1931 (which Hardy must have been aware of, but apparently never commented on) it's been known that for an area like number theory the situation is quite different: there are statements one can give within the context of the theory whose truth or falsity is undecidable from the axioms.

It was proved in the early 1960s that there are polynomial equations involving integers where it's undecidable from the axioms of arithmetic—or in effect from the formal methods of number theory—whether or not the equations have solutions. The particular examples of classes of equations where it's known that this happens are extremely complex. But from my investigations in the computational universe, I've long suspected that there are vastly simpler equations where it happens too. Over the past several decades, I've had the opportunity to poll some of the world's leading number theorists on where they think the boundary of undecidability lies. Opinions differ, but it's certainly within the realm of possibility that for example cubic equations with three variables could exhibit undecidability.

So the question then is, why should the truth of what seem like random facts of number theory even be decidable? In other words, it's perfectly possible that Ramanujan could have stated a result that simply can't be proved true or false from the axioms of arithmetic. Conceivably the Goldbach conjecture will turn out to be an example. And so could many of Ramanujan's results.

Some of Ramanujan's results have taken decades to prove—but the fact that they're provable at all is already important information. For it suggests that in a sense they're not just random facts; they're actually facts that can somehow be connected by proofs back to the underlying axioms.

And I must say that to me this tends to support the idea that Ramanujan had intuition and aesthetic criteria that in some sense captured some of the deeper principles we now know, even if he couldn't express them directly.

Automating Ramanujan

It's pretty easy to start picking mathematical statements, say at random, and then getting empirical evidence for whether they're true or not. Gödel's theorem effectively implies that you'll never know how far you'll have to go to be certain of any particular result. Sometimes it won't be far, but sometimes it may in a sense be arbitrarily far.

Ramanujan no doubt convinced himself of many of his results by what amount to empirical methods—and often it worked well. In the case of the counting of primes, however, as Hardy pointed out, things turn out to be more subtle, and results that might work up to very large numbers can eventually fail.

So let's say one looks at the space of possible mathematical statements, and picks statements that appear empirically at least to some level to be true. Now the next question: are these statements connected in any way?

Imagine one could find proofs of the statements that are true. These proofs effectively correspond to paths through a directed graph that starts with the axioms, and leads to the true results. One possibility is then that the graph is like a star—with every result being independently proved from the axioms. But another possibility is that there are many common "waypoints" in getting from the axioms to the results. And it's these waypoints that in effect represent general principles.

If there's a certain sparsity to true results, then it may be inevitable that many of them are connected through a small number of general principles. It might also be that there are results that aren't connected in this way, but these results, perhaps just because of their lack of connections, aren't considered "interesting"—and so are effectively dropped when one thinks about a particular subject.

I have to say that these considerations lead to an important question for me. I have spent many years studying what amounts to a generalization of mathematics: the behavior of arbitrary simple programs in the computational universe. And I've found that there's a huge richness of complex behavior to be seen in such programs. But I have also found evidence—not least through my Principle of Computational Equivalence—that undecidability is rife there.

But now the question is, when one looks at all that rich and complex behavior, are there in effect Ramanujan-like facts to be found there? Ultimately there will be much that can't readily be reasoned about in axiom systems like the ones in mathematics. But perhaps there are networks of facts that can be reasoned about—and that all connect to deeper principles of some kind.

We know from the idea around the Principle of Computational Equivalence that there will always be pockets of "computational reducibility": places where one will be able to identify abstract patterns and make abstract conclusions without running into undecidability. Repetitive behavior and nested behavior are two almost trivial examples. But now the question is whether among all the specific details of particular programs there are other general forms of organization to be found.

Of course, whereas repetition and nesting are seen in a great many systems, it could be that another form of organization would be seen only much more narrowly. But we don't know. And as of now, we don't really have much of a handle on finding out—at least until or unless there's a Ramanujan-like figure not for traditional mathematics but for the computational universe.

Modern Ramanujans?

Will there ever be another Ramanujan? I don't know if it's the legend of Ramanujan or just a natural feature of the way the world is set up, but for at least 30 years I've received a steady stream of letters that read a bit like the one Hardy got from Ramanujan back in 1913. Just a few months ago, for example, I received an email (from India, as it happens) with an image of a notebook listing various mathematical expressions that are numerically almost integers—very much like Ramanujan's $e^{\pi\sqrt{58}}$.

① $35\sqrt{\pi}\ln 2 = 42.9999986$

② $\frac{9\pi\,\Gamma(3/4)\,\gamma^2}{e^{6\pi/7}\cdot\gamma^{\pi/7}} \approx 1.000004248$

③ $51\ln(36\pi) - 2e^{\pi} + 21^{1/4} \approx 196.99999991695$

④ $\frac{5e^{\zeta(3)^2\times\gamma}}{\ln 10} + 137^{\pi-e} - 4\ln 2$

⑤ $46\times 3^{\pi} - \frac{\pi^3}{840}$

Are these numerical facts significant? I don't know. Wolfram|Alpha can certainly generate lots of similar facts, but without Ramanujan-like insight, it's hard to tell which, if any, are significant.

Possible closed forms:

$$35\sqrt{\pi}\log(2) \approx 42.99999862997216352 5$$

$$-e^{-\frac{5}{2}+\frac{7}{e}-e+\frac{6}{\pi}+3\pi}\,\pi^{\frac{3e}{2}-8}\,\sin^{\frac{7}{2}}(e\,\pi)\sec(e\,\pi) \approx 42.99999862997210262 5$$

$$\frac{e\,e! - 446 + 176\,e + 1133\,e^2}{72\,e} \approx 42.99999862997211082 7$$

Over the years I've received countless communications a bit like this one. Number theory is a common topic. So are relativity and gravitation theory. And particularly in recent years, AI and consciousness have been popular too. The nice thing about letters related to math is that there's typically something immediately concrete in them: some specific formula, or fact, or theorem. In Hardy's day it was hard to check such things; today it's a lot easier. But—as in the case of the almost integer above—there's then the question of whether what's being said is somehow "interesting", or whether it's just a "random uninteresting fact".

Needless to say, the definition of "interesting" isn't an easy or objective one. And in fact the issues are very much the same as Hardy faced with Ramanujan's letter. If one can see how what's being presented fits into some bigger picture—some narrative—that one understands, then one can tell whether, at least within that framework, something is "interesting". But if one doesn't have the bigger picture—or if what's being presented is just "too far out"—then one really has no way to tell if it should be considered interesting or not.

When I first started studying the behavior of simple programs, there really wasn't a context for understanding what was going on in them. The pictures I got certainly seemed visually interesting. But it wasn't clear what the bigger intellectual story was. And it took quite a few years before I'd accumulated enough empirical data to formulate hypotheses and develop principles that let one go back and see what was and wasn't interesting about the behavior I'd observed.

I've put a few decades into developing a science of the computational universe. But it's still young, and there is much left to discover—and it's a highly accessible area, with no threshold of elaborate technical knowledge. And one consequence of this is that I frequently get letters that show remarkable behavior in some particular cellular automaton or other simple program. Often I recognize the general form of the behavior, because it relates to things I've seen before, but sometimes I don't—and so I can't be sure what will or won't end up being interesting.

Back in Ramanujan's day, mathematics was a younger field—not quite as easy to enter as the study of the computational universe, but much closer than modern academic mathematics. And there were plenty of "random facts" being published: a particular type of integral done for the first time, or a new class of equations that could be solved. Many years later we would collect as many of these as we could to build them into the algorithms and knowledgebase of Mathematica and the Wolfram Language. But at the time probably the most significant aspect of their publication was the proofs that were given: the stories that explained why the results were true. Because in these proofs, there was at least the potential that concepts were introduced that could be reused elsewhere, and build up part of the fabric of mathematics.

It would take us too far afield to discuss this at length here, but there is a kind of analog in the study of the computational universe: the methodology for computer experiments. Just as a proof can contain elements that define a general methodology for getting a mathematical result, so the particular methods of search, visualization or analysis can define something in computer experiments that is general and reusable, and can potentially give an indication of some underlying idea or principle.

And so, a bit like many of the mathematics journals of Ramanujan's day, I've tried to provide a journal and a forum where specific results about the computational universe can be reported—though there is much more that could be done along these lines.

When a letter one receives contains definite mathematics, in mathematical notation, there is at least something concrete one can understand in it. But plenty of things can't usefully be formulated in mathematical notation. And too often, unfortunately, letters are in plain English (or worse, for me, other languages) and it's almost impossible for me to tell what they're trying to say. But now there's something much better that people increasingly do: formulate things in the Wolfram Language. And in that form, I'm always able to tell what someone is trying to say—although I still may not know if it's significant or not.

Over the years, I've been introduced to many interesting people through letters they've sent. Often they'll come to our Summer School, or publish something in one of our various channels. I have no story (yet) as dramatic as Hardy and Ramanujan. But it's wonderful that it's possible to connect with people in this way, particularly in their formative years. And I can't forget that a long time ago, I was a 14-year-old who mailed papers about the research I'd done to physicists around the world...

What If Ramanujan Had Mathematica?

Ramanujan did his calculations by hand—with chalk on slate, or later pencil on paper. Today with Mathematica and the Wolfram Language we have immensely more powerful tools with which to do experiments and make discoveries in mathematics (not to mention the computational universe in general).

It's fun to imagine what Ramanujan would have done with these modern tools. I rather think he would have been quite an adventurer—going out into the mathematical universe and finding all sorts of strange and wonderful things, then using his intuition and aesthetic sense to see what fits together and what to study further.

Ramanujan unquestionably had remarkable skills. But I think the first step to following in his footsteps is just to be adventurous: not to stay in the comfort of well-established mathematical theories, but instead to go out into the wider mathematical universe and start finding—experimentally—what's true.

It's taken the better part of a century for many of Ramanujan's discoveries to be fitted into a broader and more abstract context. But one of the great inspirations that Ramanujan gives us is that it's possible with the right sense to make great progress even before the broader context has been understood. And I for one hope that many more people will take advantage of the tools we have today to follow Ramanujan's lead and make great discoveries in experimental mathematics—whether they announce them in unexpected letters or not.

Solomon Golomb

May 25, 2016

The Most-Used Mathematical Algorithm Idea in History

An octillion. A billion billion billion. That's a fairly conservative estimate of the number of times a cellphone or other device somewhere in the world has generated a bit using a maximum-length linear-feedback shift register sequence. It's probably the single most-used mathematical algorithm idea in history. And the main originator of this idea was Solomon Golomb, who died on May 1—and whom I knew for 35 years.

Solomon Golomb's classic book *Shift Register Sequences*, published in 1967—based on his work in the 1950s—went out of print long ago. But its content lives on in pretty much every modern communications system. Read the specifications for 3G, LTE, Wi-Fi, Bluetooth, or for that matter GPS, and you'll find mentions of polynomials that determine the shift register sequences these systems use to encode the data they send. Solomon Golomb is the person who figured out how to construct all these polynomials.

He also was in charge when radar was first used to find the distance to Venus, and of working out how to encode images to be sent from Mars. He introduced the world to what he called polyominoes, which later inspired Tetris ("tetromino tennis"). He created and solved countless math and wordplay puzzles. And—as I learned about 20 years ago—he came very close to discovering my all-time-favorite rule 30 cellular automaton all the way back in 1959, the year I was born.

How I Met Sol Golomb

Most of the scientists and mathematicians I know I met first through professional connections. But not Sol Golomb. It was 1981, and I was at Caltech, a 21-year-old physicist who'd just received some media attention from being the youngest in the first batch of MacArthur award recipients. I get a knock at my office door—and a young woman is there. Already this was unusual, because in those days there were hopelessly few women to be found around a theoretical high-energy physics group. I was a sheltered Englishman who'd been in California a couple of years, but hadn't really ventured outside the university—and was ill prepared for the burst of Southern Californian energy that dropped in to see me that day. She introduced herself as Astrid, and said that she'd been visiting Oxford and knew someone I'd been at kindergarten with. She explained that she had a personal mandate to collect interesting acquaintances around the Pasadena area. I think she considered me a difficult case, but persisted nevertheless. And one day when I tried to explain something about the work I was doing she said, "You should meet my father. He's a bit old, but he's still as sharp as a tack." And so it was that Astrid Golomb, oldest daughter of Sol Golomb, introduced me to Sol Golomb.

The Golombs lived in a house perched in the hills near Pasadena. I learned that they had two daughters—Astrid, a little older than me, an aspiring Hollywood person, and Beatrice, about my age, a high-powered science type. The Golomb sisters often had parties, usually at their family's house. There were themes, like the flamingoes & hedgehogs croquet garden party ("recognition will be given to the person who appears most appropriately attired"), or the Stonehenge party with instructions written using runes. The parties had an interesting cross-section of young and not-so-young people, including various local luminaries. And always there, hanging back a little, was Sol Golomb, a small man with a large beard and a certain elf-like quality to him, typically wearing a dark suit coat.

I gradually learned a little about Sol Golomb. That he was involved in "information theory". That he worked at USC (the University of Southern California). That he had various unspecified but apparently high-level government and other connections. I'd heard of shift registers, but didn't really know anything much about them.

Then in the fall of 1982, I visited Bell Labs in New Jersey and gave a talk about my latest results on cellular automata. One topic I discussed was what I called "additive" or "linear" cellular automata—and their behavior with limited numbers of cells. Whenever a cellular automaton has a limited number of cells, it's inevitable that its behavior will eventually repeat. But as the size increases, the maximum repetition period—say for the rule 90 additive cellular automaton—bounces around seemingly quite randomly: 1, 1, 3, 2, 7, 1, 7, 6, 31, 4, 63, ... A few days before my talk, however, I'd noticed that these periods actually seemed to follow a formula that depended on things like the prime factorization of the number of cells. But when I mentioned this during the talk, someone at the back put up their hand and asked, "Do you know if it works for the case n=37?" My experiments hadn't gotten as far as the size-37 case yet, so I didn't know. But why would someone ask that?

The person who asked turned out to be a certain Andrew Odlyzko, a number theorist at Bell Labs. I asked him, "What on earth makes you think there might be something special about n=37?" "Well," he said, "I think what you're doing is related to the theory of linear-feedback shift registers," and he suggested that I look at Sol Golomb's book ("Oh yes," I said, "I know his daughters..."). Andrew was indeed correct: there is a very elegant theory of additive cellular automata based on polynomials that is similar to the theory Sol developed for linear-feedback shift registers. Andrew and I ended up writing a now-rather-well-cited paper about it (it's interesting because it's a rare case where traditional mathematical methods let one say things about nontrivial cellular automaton behavior). And for me, a side effect was that I learned something about what the somewhat mysterious Sol Golomb actually did. (Remember, this was before the web, so one couldn't just instantly look everything up.)

The Story of Sol Golomb

Solomon Golomb was born in Baltimore, Maryland in 1932. His family came from Lithuania. His grandfather had been a rabbi; his father moved to the US when he was young, and got a master's degree in math before switching to medieval Jewish philosophy and also becoming a rabbi. Sol's mother came from a prominent Russian family that had made boots for the Tsar's army and then ran a bank. Sol did well in school, notably being a force in the local debating scene. Encouraged by his father, he developed an interest in mathematics, publishing a problem he invented about primes when he was 17. After high school, Sol enrolled at Johns Hopkins University to study math, narrowly avoiding a quota on Jewish students by promising he wouldn't switch to medicine—and took twice the usual course load, graduating in 1951 after half the usual time.

From there he would go to Harvard for graduate school in math. But first he took a summer job at the Glenn L. Martin Company, an aerospace firm founded in 1912 that had moved to Baltimore from Los Angeles in the 1920s and mostly become a defense contractor—and that would eventually merge into Lockheed Martin. At Harvard, Sol specialized in number theory, and in particular in questions about characterizations of sets of prime numbers. But every summer he would return to the Martin Company. As he later described it, he found that at Harvard "the question of whether anything that was taught or studied in the mathematics department had any practical applications could not even be asked, let alone discussed". But at the Martin Company, he discovered that the pure mathematics he knew—even about primes and things—did indeed have practical applications, and very interesting ones, especially to shift registers.

The first summer he was at the Martin Company, Sol was assigned to a control theory group. But by his second summer, he'd been put in a group studying communications. And in June 1954 it so happened that his supervisor had just gone to a conference where he'd heard about strange behavior observed in linear-feedback shift registers (he called them "tapped delay lines with feedback")—and he asked Sol if he could investigate. It didn't take Sol long to realize that what was going on could be very elegantly studied using the pure mathematics he knew about polynomials over finite fields. Over the year that followed, he split his time between graduate school at Harvard and consulting for the Martin Company, and in June 1955 he wrote his final report, "Sequences with Randomness Properties"—which would basically become the foundational document of the theory of shift register sequences.

Sol liked math puzzles, and in the process of thinking about a puzzle involving arranging dominoes on a checkerboard, he ended up inventing what he called "polyominoes". He gave a talk about them in November 1953 at the Harvard Mathematics Club, published a paper about them (his first research publication), won a Harvard math prize for his work on them, and, as he later said, then "found [himself] irrevocably committed to their care and feeding" for the rest of his life.

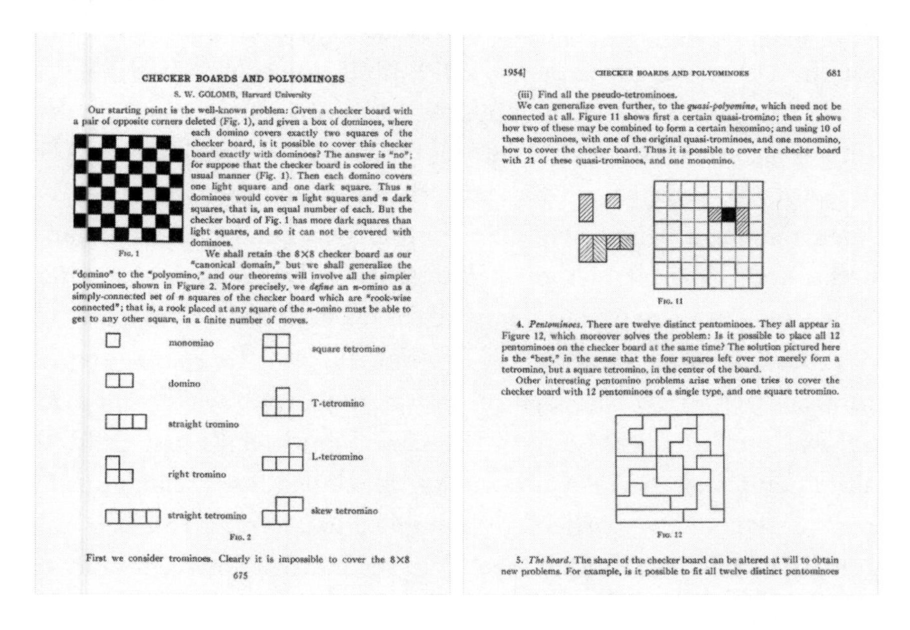
CHECKER BOARDS AND POLYOMINOES

S. W. GOLOMB, Harvard University

Our starting point is the well-known problem: Given a checker board with a pair of opposite corners deleted (Fig. 1), and given a box of dominoes, where each domino covers exactly two squares of the checker board, is it possible to cover this checker board exactly with dominoes? The answer is "no"; for suppose that the checker board is colored in the usual manner (Fig. 1). Then each domino covers one light square and one dark square. Thus n dominoes would cover n light squares and n dark squares, that is, an equal number of each. But the checker board of Fig. 1 has more dark squares than light squares, and so it can not be covered with dominoes.

Fig. 1

We shall retain the 8×8 checker board as our "canonical domain," but we shall generalize the "domino" to the "polyomino," and our theorems will involve all the simpler polyominoes, shown in Figure 2. More precisely, we *define* an n-omino as a simply-connected set of n squares of the checker board which are "rook-wise connected"; that is, a rook placed at any square of the n-omino must be able to get to any other square, in a finite number of moves.

Fig. 2

First we consider trominoes. Clearly it is impossible to cover the 8×8

675

1954] CHECKER BOARDS AND POLYOMINOES 681

(iii) Find all the pseudo-tetrominoes.

We can generalize even further, to the *quasi-polyomino*, which need not be connected at all. Figure 11 shows first a certain quasi-tromino; then it shows how two of these may be combined to form a certain hexomino; and using 10 of these hexominoes, with one of the original quasi-trominoes, and one monomino, how to cover the checker board. Thus it is possible to cover the checker board with 21 of these quasi-trominoes, and one monomino.

Fig. 11

4. *Pentominoes.* There are twelve distinct pentominoes. They all appear in Figure 12, which moreover solves the problem: Is it possible to place all 12 pentominoes on the checker board at the same time? The solution pictured here is the "best," in the sense that the four squares left over not merely form a tetromino, but a square tetromino, in the center of the board.

Other interesting pentomino problems arise when one tries to cover the checker board with 12 pentominoes of a single type, and one square tetromino.

Fig. 12

5. *The board.* The shape of the checker board can be altered at will to obtain new problems. For example, is it possible to fit all twelve distinct pentominoes

In June 1955, Sol went to spend a year at the University of Oslo on a Fulbright Fellowship—partly so he could work with some distinguished number theorists there, and partly so he could add Norwegian, Swedish and Danish (and some runic scripts) to his collection of language skills. While he was there, he finished a long paper on prime numbers, but also spent time traveling around Scandinavia, and in Denmark met a young woman named Bo (Bodil Rygaard)—who came from a large family in a rural area mostly known for its peat moss, but had managed to get into university and was studying philosophy. Sol and Bo apparently hit it off, and within months, they were married.

When they returned to the US in July 1956, Sol interviewed in a few places, then accepted a job at JPL—the Jet Propulsion Lab that had spun off from Caltech, initially to do military work. Sol was assigned to the Communications Research Group, as a Senior Research Engineer. It was a time when the people at JPL were eager to try launching a satellite. At first, the government wouldn't let them do it, fearing it would be viewed as a military act. But that all changed in October 1957 when the Soviet Union launched Sputnik, ostensibly as part of the International Geophysical Year. Amazingly, it took only 3 months for the US to launch Explorer 1. JPL built much of it, and Sol's lab (where he had technicians building electronic implementations of shift registers) was diverted into doing things like making radiation detectors (including, as it happens, the ones that discovered the Van Allen radiation belts)—while Sol himself worked on using radar to determine the orbit of the satellite when it was launched, taking a little time out to go back to Harvard for his final PhD exam.

It was a time of great energy around JPL and the space program. In May 1958 a new Information Processing Group was formed, and Sol was put in charge—and in the same month, Sol's first child, the aforementioned Astrid, was born. Sol continued his research on shift register sequences—particularly as applied to jamming-resistant radio control of missiles. In May 1959, Sol's second child arrived—and was named Beatrice, forming a nice A, B sequence. In the fall of 1959, Sol took a sabbatical at MIT, where he got to know Claude Shannon and a number of other MIT luminaries, and got involved in information theory and the theory of algebraic codes.

As it happens, he'd already done some work on coding theory—in the area of biology. The digital nature of DNA had been discovered by Jim Watson and Francis Crick in 1953, but it wasn't yet clear just how sequences of the four possible base pairs encoded the 20 amino acids. In 1956, Max Delbrück—Jim Watson's former postdoc advisor at Caltech—asked around at JPL if anyone could figure it out. Sol and two colleagues analyzed an idea of Francis Crick's and came up with "comma-free codes" in which overlapping triples of base pairs could encode amino acids. The analysis showed that exactly 20 amino acids could be encoded this way. It seemed like an amazing explanation of what was seen—but unfortunately it isn't how biology actually works (biology uses a more straightforward encoding, where some of the 64 possible triples just don't represent anything).

In addition to biology, Sol was also pulled into physics. His shift register sequences were useful for doing range finding with radar (much as they're used now in GPS), and at Sol's suggestion, he was put in charge of trying to use them to find the distance to Venus. And so it was that in early 1961—when the Sun, Venus, and Earth were in alignment—Sol's team used the 85-foot Goldstone radio dish in the Mojave Desert to bounce a radar signal off Venus, and dramatically improve our knowledge of the Earth-Venus and Earth-Sun distances.

With his interest in languages, coding and space, it was inevitable that Sol would get involved in the question of communications with extraterrestrials. In 1961 he wrote a paper for the Air Force entitled "A Short Primer for Extraterrestrial Linguistics", and over the next several years wrote several papers on the subject for broader audiences. He said that "There are two questions involved in communication with Extraterrestrials. One is the mechanical issue of discovering a mutually acceptable channel. The other is the more philosophical problem (semantic, ethic, and metaphysical) of the proper subject matter for discourse. In simpler terms, we first require a common language, and then we must think of something clever to say." He continued, with a touch of his characteristic humor: "Naturally, we must not risk telling too much until we know whether the Extraterrestrials' intentions toward us are honorable. The Government will undoubtedly set up a Cosmic Intelligence Agency (CIA) to monitor Extraterrestrial Intelligence. Extreme security precautions will be strictly observed. As H. G. Wells once pointed out [or was it an episode of *The Twilight Zone*?], even if the Aliens tell us in all truthfulness that their only intention is 'to serve mankind,' we must endeavor to ascertain whether they wish to serve us baked or fried."

While at JPL, Sol had also been teaching some classes at the nearby universities: Caltech, USC and UCLA. In the fall of 1962, following some changes at JPL—and perhaps because he wanted to spend more time with his young children—he decided to become a full-time professor. He got offers from all three schools. He wanted to go somewhere where he could "make a difference". He was told that at Caltech "no one has any influence if they don't at least have a Nobel Prize", while at UCLA "the UC bureaucracy is such that no one ever has any ability to affect anything". The result was that—despite its much-inferior reputation at the time—Sol chose USC. He went there in the spring of 1963 as a Professor of Electrical Engineering—and ended up staying for 53 years.

Shift Registers

Before going on with the story of Sol's life, I should explain what a linear-feedback shift register (LFSR) actually is. The basic idea is simple. Imagine a row of squares, each containing either 1 or 0 (say, black or white). In a pure shift register all that happens is that at each step all values shift one position to the left. The leftmost value is lost, and a new value is "shifted in" from the right. The idea of a feedback shift register is that the value that's shifted in is determined (or "fed back") from values at other positions in the shift register. In a linear-feedback shift register, the values from "taps" at particular positions in the register are combined by being added mod 2 (so that 1⊕1=0 instead of 2), or equivalently XOR'ed ("exclusive or", true if either is true, but not both).

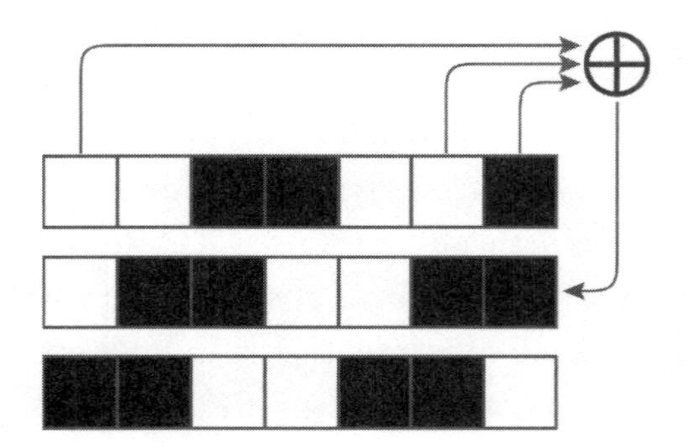

If one runs this for a while, here's what happens:

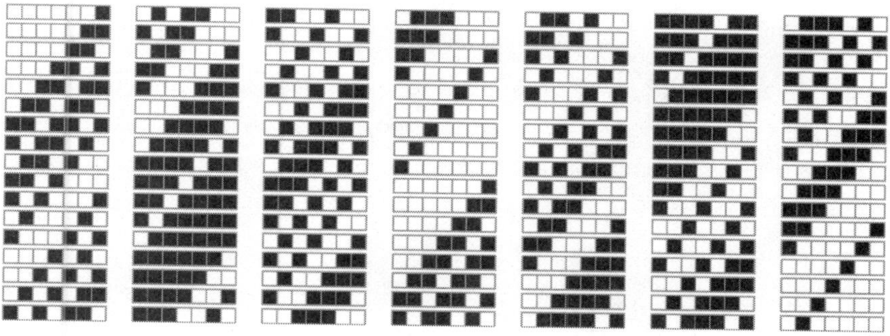

Obviously the shift register is always shifting bits to the left. And it has a very simple rule for how bits should be added at the right. But if one looks at the sequence of these bits, it seems rather random—though, as the picture shows, it does eventually repeat. What Sol Golomb did was to find an elegant mathematical way to analyze such sequences, and how they repeat.

If a shift register has size n, then it has 2^n possible states altogether (corresponding to all possible sequences of 0s and 1s of length n). Since the rules for the shift register are deterministic, any given state must always go to the same next state. And that means the maximum possible number of steps the shift register could conceivably go through before it repeats is 2^n (actually, it's 2^n-1, because the state with all 0s can't evolve into anything else).

In the example above, the shift register is of size 7, and it turns out to repeat after exactly $2^7-1 = 127$ steps. But which shift registers—with which particular arrangements of taps—will produce sequences with maximal lengths? This is the first question Sol Golomb set out to investigate in the summer of 1954. His answer was simple and elegant.

The shift register above has taps at positions 7, 6 and 1. Sol represented this algebraically, using the polynomial x^7+x^6+1. Then what he showed was that the sequence that would be generated would be of maximal length if this polynomial is "irreducible modulo 2", so that it can't be factored, making it sort of the analog of a prime among polynomials—as well as having some other properties that make it a so-called "primitive polynomial". Nowadays, with Mathematica and the Wolfram Language, it's easy to test things like this:

```
In[1]:= IrreduciblePolynomialQ[x^7 + x^6 + 1, Modulus → 2]
Out[1]= True
```

Back in 1954, Sol had to do all this by hand, but came up with a fairly long table of irreducible polynomials corresponding to shift registers that give maximal length sequences:

Table III-5. Irreducible polynomials modulo 2, through degree 11, with their periods.

Degree (bold) & polynomial*	Period
1	
2	—
3	1
2	
7	3
3	
13	7
15	7
4	
23	15
31	15
37	5
5	
45	31
51	31
57	31
67	31
73	31
75	31
6	
103	63
111	9
127	21
133	63
141	63
147	63
155	63
163	63
165	21
7	
203	127

Degree (bold) & polynomial	Period
211	127
217	127
221	127
235	127
247	127
253	127
271	127
277	127
301	127
313	127
323	127
325	127
345	127
357	127
361	127
367	127
375	127
8	
433	51
435	255
453	255
455	255
471	17
477	85
515	255
537	255
543	255
545	255
551	255
561	255
567	85
573	85
607	255

Degree (bold) & polynomial	Period
613	85
615	255
637	51
643	85
651	255
661	51
675	85
703	255
717	255
727	17
735	85
747	255
763	51
765	255
771	85
9	
1003	73
1021	511
1027	73
1033	511
1041	511
1055	511
1063	511
1113	73
1137	511
1145	73
1151	511
1157	511
1167	511
1175	511
1207	511
1225	511
1231	73

* If $f(x) = \sum_{i=0}^{n} c_i x^i$, the table entry is $\sum_{i=0}^{n} c_i 2^i$ written to the base 8. Thus $x^5 + x^3 + x^2 + x + 1$ becomes binary 101. 111 which is octal "57."

Table III-5 (Cont'd).

Degree (bold) & polynomial	Period	Degree (bold) & polynomial	Period	Degree (bold) & polynomial	Period
1243	511	1773	511	2547	341
1245	511			2553	1023
1257	511	**10**		2605	1023
1267	511	2011	1023	2617	1023
1275	511	2017	341	2627	1023
1317	511	2033	1023	2633	341
1321	511	2035	341	2641	1023
1333	511	2047	1023	2653	341
1365	511	2055	1023	2671	341
1371	511	2065	93	2701	341
1401	73	2107	341	2707	1023
1423	511	2123	341	2745	1023
1425	511	2143	341	2767	1023
1437	511	2145	1023	2773	1033
1443	511	2157	1023	3023	1023
1461	511	2201	1023	3025	1023
1473	511	2213	1023	3043	33
1511	73	2231	341	3045	1023
1517	511	2251	33	3061	341
1533	511	2257	341	3067	1023
1541	511	2305	1023	3103	1023
1553	511	2311	341	3117	1023
1555	511	2327	1023	3121	341
1563	511	2347	1023	3133	1023
1577	511	2355	341	3171	1023
1605	511	2363	1023	3177	1023
1617	511	2377	1023	3205	93
1641	73	2413	93	3211	1023
1665	511	2415	1023	3247	93
1671	511	2431	1023	3255	341
1707	511	2437	341	3265	1023
1713	511	2443	1023	3277	341
1715	511	2461	1023	3301	1023
1725	511	2475	1023	3315	341
1731	511	2503	1023	3323	1023
1743	511	2527	1023	3337	1023
1751	511	2541	93	3367	341

The Prehistory of Shift Registers

The idea of maintaining short-term memory by having "delay lines" that circulate digital pulses (say in an actual column of mercury) goes back to the earliest days of electronic computers. By the late 1940s such delay lines were routinely being implemented purely digitally, using sequences of vacuum tubes, and were being called "shift registers". It's not clear when the first feedback shift registers were built. Perhaps it was at the end of the 1940s. But it's still shrouded in mystery—because the first place they seem to have been used was in military cryptography.

The basic idea of cryptography is to take meaningful messages, and then randomize them so they can't be recognized, but in such a way that the randomization can always be reversed if you know the key that was used to create it. So-called stream ciphers work by generating long sequences of seemingly random bits, then combining these with some representation of the message—then decoding by having the receiver independently generate the same sequence of seemingly random bits, and "backing this out" of the encoded message received.

Linear-feedback shift registers seem at first to have been prized for cryptography because of their long repetition periods. As it turns out, the mathematical analysis Sol used to find things like these periods also makes clear that such shift registers aren't good for secure cryptography. But in the early days, they seemed pretty good—particularly compared to, say, successive rotor positions in an Enigma machine—and there's been a persistent rumor that, for example, Soviet military cryptosystems were long based on them.

Back in 2001, when I was working on history notes for my book *A New Kind of Science*, I had a long phone conversation with Sol about shift registers. Sol told me that when he started out, he didn't know anything about cryptographic work on shift registers. He said that people at Bell Labs, Lincoln Labs and JPL had also started working on shift registers around the same time he did—though perhaps through knowing more pure mathematics, he managed to get further than they did, and in the end his 1955 report basically defined the field.

Over the years that followed, Sol gradually heard about various precursors of his work in the pure mathematical literature. Way back in the year 1202 Fibonacci was already talking about what are now called Fibonacci numbers—and which are generated by a recurrence relation that can be thought of as an analog of a linear-feedback shift register, but working with arbitrary integers rather than 0s and 1s. There was a little work on recurrences with 0s and 1s done in the early 1900s, but the first large-scale study seems to have been by Øystein Ore, who, curiously, came from the University of Oslo, though was by then at Yale. Ore had a student named Marshall Hall—who Sol told me he knew had consulted for the predecessor of the National Security Agency in the late 1940s—possibly about shift registers. But whatever he may have done was kept secret, and so it fell to Sol to discover and publish the story of linear-feedback shift registers—even though Sol did dedicate his 1967 book on shift registers to Marshall Hall.

What Are Shift Register Sequences Good For?

Over the years I've noticed the principle that systems defined by sufficiently simple rules always eventually end up having lots of applications. Shift registers follow this principle in spades. And for example modern hardware (and software) systems are bristling with shift registers: a typical cellphone probably has a dozen or two, implemented usually in hardware but sometimes in software. (When I say "shift register" here, I mean linear-feedback shift register, or LFSR.)

Most of the time, the shift registers that are used are ones that give maximum-length sequences (otherwise known as "m-sequences"). And the reasons they're used are typically related to some very special properties that Sol discovered about them. One basic property they always have is that they contain the same total number of 0s and 1s (actually, there's always exactly one extra 1). Sol then showed that they also have the same number of 00s, 01s, 10s and 11s—and the same holds for larger blocks too. This "balance" property is on its own already very useful, for example if one's trying to efficiently test all possible bit patterns as input to a circuit.

But Sol discovered another, even more important property. Replace each 0 in a sequence by −1, then imagine multiplying each element in a shifted version of the sequence by the corresponding element in the original. What Sol showed is that if one adds up these products, they'll always sum to zero, except when there's no shift at all. Said more technically, he showed that the sequence has no correlation with shifted versions of itself.

Both this and the balance property will be approximately true for any sufficiently long random sequence of 0s and 1s. But the surprising thing about maximum-length shift register sequences is that these properties are always exactly true. The sequences in a sense have some of the signatures of randomness—but in a very perfect way, made possible by the fact that they're not random at all, but instead have a very definite, organized structure.

It's this structure that makes linear-feedback shift registers ultimately not suitable for strong cryptography. But they're great for basic "scrambling" and "cheap cryptography"—and they're used all over the place for these purposes. A very common objective is just to "whiten" (as in "white noise") a signal. It's pretty common to want to transmit data that's got long sequences of 0s in it. But the electronics that pick these up can get confused if they see what amounts to "silence" for too long. One can avoid the problem by scrambling the original data by combining it with a shift register sequence, so there's always some kind of "chattering" going on. And that's indeed what's done in Wi-Fi, Bluetooth, USB, digital TV, Ethernet and lots of other places.

It's often a nice side effect that the shift register scrambling makes the signal harder to decode—and this is sometimes used to provide at least some level of security. (DVDs use a combination of a size-16 and a size-24 shift register to attempt to encode their data; many GSM phones use a combination of three shift registers to encode all their signals, in a way that was at first secret.)

GPS makes crucial use of shift register sequences too. Each GPS satellite continuously transmits a shift register sequence (from a size-10 shift register, as it happens). A receiver can tell at exactly what time a signal it's just received was transmitted from a particular satellite by seeing what part of the sequence it got. And by comparing delay times from different satellites, the receiver can triangulate its position. (There's also a precision mode of GPS, that uses a size-1024 shift register.)

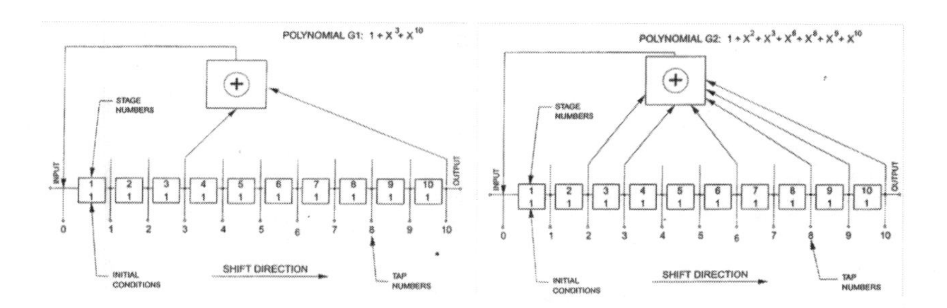

A quite different use of shift registers is for error detection. Say one's transmitting a block of bits, but each one has a small probability of error. A simple way to let one check for a single error is to include a "parity bit" that says whether there should be an odd or even number of 1s in the block of bits. There are generalizations of this called CRCs (cyclic redundancy checks) that can check for a larger number of errors—and that are computed essentially by feeding one's data into none other than a linear-feedback shift register. (There are also error-correcting codes that let one not only detect but also correct a certain number of errors, and some of these, too, can be computed with shift register sequences—and in fact Sol Golomb used a version of these called Reed–Solomon codes to design the video encoding for Mars spacecraft.)

The list of uses for shift register sequences goes on and on. A fairly exotic example—more popular in the past than now—was to use shift register sequences to jitter the clock in a computer to spread out the frequency at which the CPU would potentially generate radio interference ("select Enable Spread Spectrum in the BIOS").

One of the single most prominent uses of shift register sequences is in cellphones, for what's called CDMA (Code Division Multiple Access). Cellphones got their name because they operate in "cells", with all phones in a given cell being connected to a particular tower. But how do different cellphones in a cell not interfere with each other? In the first systems, each phone just negotiated with the tower to use a slightly different frequency. Later, they used different time slices (TDMA, or Time Division Multiple Access). But CDMA uses maximum-length shift register sequences to provide a clever alternative.

The idea is to have all phones essentially operate on the same frequency, but to have each phone encode its signal using (in the simplest case) a differently shifted version of a shift register sequence. And because of Sol's mathematical results, these differently shifted versions have no correlation—so the cellphone signals don't interfere. And this is how, for example, most 3G cellphone networks operate.

Sol created the mathematics for this, but he also brought some of the key people together. Back in 1959, he'd gotten to know a certain Irwin Jacobs, who'd recently gotten a PhD at MIT. Meanwhile, he knew Andy Viterbi, who worked at JPL. Sol introduced the two of them—and by 1968 they'd formed a company called Linkabit which did work on coding systems, mostly for the military.

Linkabit had many spinoffs and descendents, and in 1985 Jacobs and Viterbi started a new company called Qualcomm. It didn't immediately do especially well, but by the early 1990s it began a meteoric rise when it started making the components to deploy CDMA in cellphones—and in 1999 Sol became the "Viterbi Professor of Communications" at USC.

Where Are There Shift Registers?

It's sort of amazing that—although most people have never heard of them—shift register sequences are actually used in one way or another almost whenever bits are moved around in modern communication systems, computers and elsewhere. It's quite confusing sometimes, because there are lots of things with different names and acronyms that all turn out to be linear-feedback shift register sequences (PN, pseudonoise, M-, FSR, LFSR sequences, spread spectrum communications, MLS, SRS, PRBS, ...).

If one looks at cellphones, shift register sequence usage has gone up and down over the years. 2G networks are based on TDMA, so don't use shift register sequences to encode their data—but still often use CRCs to validate blocks of data. 3G networks are big users of CDMA—so there are shift register sequences involved in pretty much every bit that's transmitted. 4G networks typically use a combination of time and frequency slots which don't directly involve shift register sequences—though there are still CRCs used, for example to deal with data integrity when frequency windows overlap. 5G is designed to be more elaborate—with large arrays of antennas dynamically adapting to use optimal time and frequency slots. But half their channels are typically allocated to "pilot signals" that are used to infer the local radio environment—and work by transmitting none other than shift register sequences.

Throughout most kinds of electronics it's common to want to use the highest data rates and the lowest powers that still get bits transmitted correctly above the "noise floor". And typically the way one pushes to the edge is to do automatic error detection—using CRCs and therefore shift register sequences. And in fact pretty much every kind of bus (PCIe, SATA, etc.) inside a computer does this: whether it's connecting parts of CPUs, getting data off devices, or connecting to a display with HDMI. And on disks and in memory, for example, CRCs and other shift-register-sequence-based codes are pretty much universally used to operate at the highest possible rates and densities.

Shift registers are so ubiquitous, it's a little difficult to estimate just how many of them are in use, and how many bits are being generated by them. There are perhaps 10 billion computers, slightly fewer cellphones, and an increasing number of billions of embedded and IoT ("Internet of Things") devices. (Even many of the billion cars in the world, for example, have at least 10 microprocessors in them.)

At what rate are the shift registers running? Here, again, things are complicated. In communications systems, for example, there's a basic carrier frequency—usually in the GHz range—and then there's what's called a "chipping rate" (or, confusingly, "chip rate") that says how fast something like CDMA is done, and this is usually in the MHz range. On the other hand, in buses inside computers, or in connections to a display, all the data is going through shift registers, at the full data rate, which is well into the GHz range.

So it seems safe to estimate that there are at least 10 billion communications links, running for at least 1/10 billion seconds (which is 3 years), that use at least 1 billion bits from a shift register every second—meaning that to date Sol's algorithm has been used at least an octillion times.

Is it really the most-used mathematical algorithm idea in history? I think so. I suspect the main potential competition would be from arithmetic operations. These days processors are doing perhaps a trillion arithmetic operations per second—and such operations are needed for pretty much every bit that's generated by a computer. But how is arithmetic done? At some level it's just a digital electronics implementation of the way people have done arithmetic forever.

But there are some wrinkles—some "algorithmic ideas"—though they're quite obscure, except to microprocessor designers. Just as when Babbage was making his Difference Engine, carries are a big nuisance in doing arithmetic. (One can actually think of a linear-feedback shift register as being a system that does something like arithmetic, but doesn't do carries.) There are "carry propagation trees" that optimize carrying. There are also little tricks ("Booth encoding", "Wallace trees", etc.) that reduce the number of bit operations needed to do the innards of arithmetic. But unlike with LFSRs, there doesn't seem to be one algorithmic idea that's universally used—and so I think it's still likely that Sol's maximum-length LFSR sequence idea is the winner for most-used.

Cellular Automata and Nonlinear Shift Registers

Even though it's not obvious at first, it turns out there's a very close relationship between feedback shift registers and something I've spent many years studying: cellular automata. The basic setup for a feedback shift register involves computing one bit at a time. In a cellular automaton, one has a line of cells, and at each step all the cells are updated in parallel, based on a rule that depends, say, on the values of their nearest neighbors.

To see how these are related, think about running a feedback shift register of size n, but displaying its state only every n steps—in other words, letting all the bits be rewritten before one displays again. If one displays every step of a linear-feedback shift register (here with two taps next to each other), as in the first two panels below, nothing much happens at each step, except that things shift to the left. But if one makes a compressed picture, showing only every n steps, suddenly a pattern emerges.

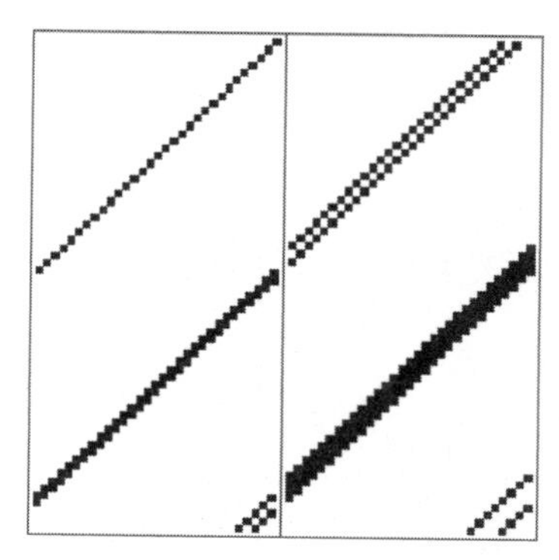

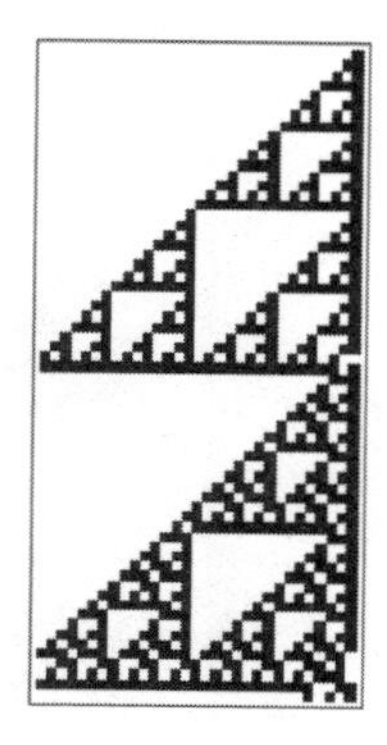

It's a nested pattern, and it's very close to being the exact same pattern that one gets with a cellular automaton that takes a cell and its neighbor, and adds them mod 2 (or XORs them). Here's what happens with that cellular automaton, if one arranges its cells so they're in a circle of the same size as the shift register above:

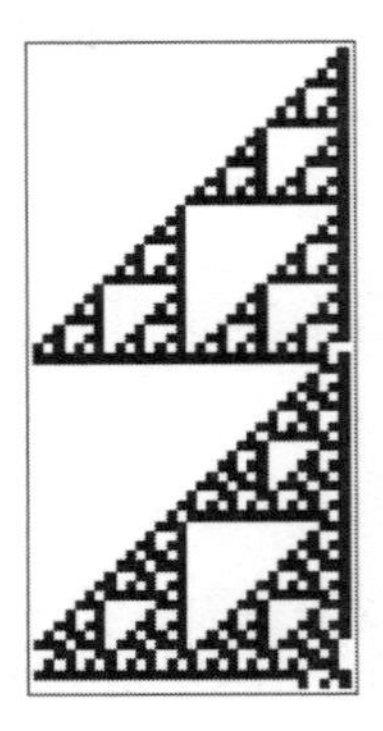

At the beginning, the cellular automaton and shift register patterns are exactly the same—though when they "hit the edge" they become slightly different because the edges are handled differently. But looking at these pictures it becomes less surprising that the math of shift registers should be relevant to cellular automata. And seeing the regularity of the nested patterns makes it clearer why there might be an elegant mathematical theory of shift registers in the first place.

Typical shift registers used in practice don't tend to make such obviously regular patterns, though. Here are a few examples of shift registers that yield maximum-length sequences. When one's doing math, like Sol did, it's very much the same story as for the case of obvious nesting. But here the fact that the taps are far apart makes things get mixed up, leaving no obvious visual trace of nesting.

So how broad is the correspondence between shift registers and cellular automata? In cellular automata the rules for generating new values of cells can be anything one wants. In linear-feedback shift registers, however, they always have to be based on adding mod 2 (or XOR'ing). But that's what the "linear" part of "linear-feedback shift register" means. And it's also in principle possible to have nonlinear-feedback shift registers (NFSRs) that use whatever rule one wants for combining values.

And in fact, once Sol had worked out his theory for linear-feedback shift registers, he started in on the nonlinear case. When he arrived at JPL in 1956 he got an actual lab, complete with racks of little electronic modules. Sol told me each module was about the size of a cigarette pack—and was built from a Bell Labs design to perform a particular logic operation (AND, OR, NOT, ...). The modules could be strung together to implement whatever nonlinear-feedback shift register one wanted, and they ran pretty fast—producing about a million bits per second. (Sol told me that someone tried doing the same thing with a general-purpose computer—and what took 1 second with the custom hardware modules took 6 weeks on the general-purpose computer.)

When Sol had looked at linear-feedback shift registers, the first big thing he'd managed to understand was their repetition periods. And with nonlinear ones he put most of his effort into trying to understand the same thing. He collected all sorts of experimental data. He told me he even tested sequences of length 2^{45}—which must have taken a year. He made summaries, like the one below (notice the visualizations of sequences, shown as oscilloscope-like traces). But he never managed to come up with any kind of general theory as he had with linear-feedback shift registers.

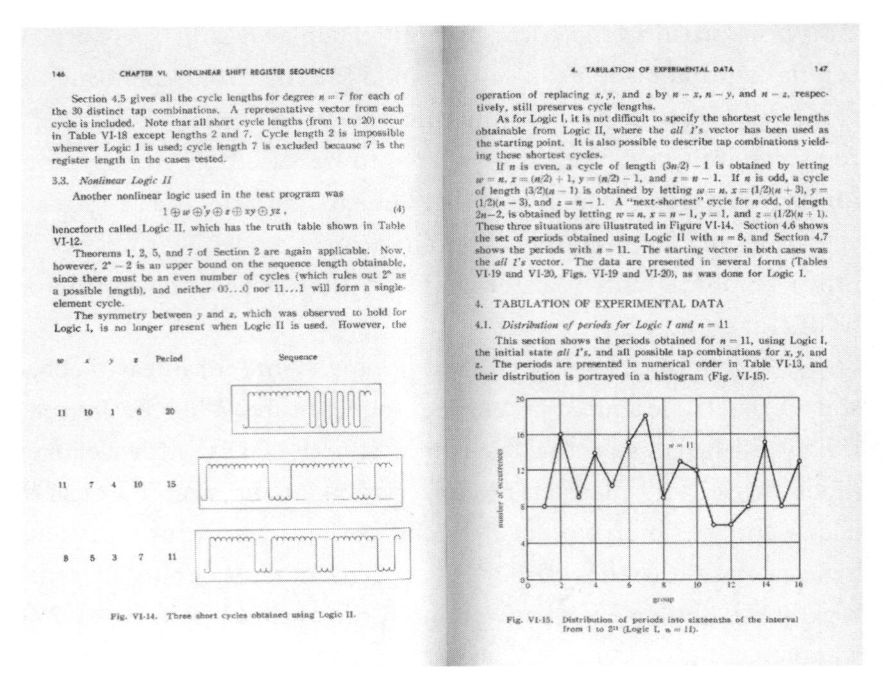
146 CHAPTER VI. NONLINEAR SHIFT REGISTER SEQUENCES

Section 4.5 gives all the cycle lengths for degree $n = 7$ for each of the 30 distinct tap combinations. A representative vector from each cycle is included. Note that all short cycle lengths (from 1 to 20) occur in Table VI-18 except lengths 2 and 7. Cycle length 2 is impossible whenever Logic I is used; cycle length 7 is excluded because 7 is the register length in the cases tested.

3.3. *Nonlinear Logic II*

Another nonlinear logic used in the test program was

$$1 \oplus w \oplus y \oplus z \oplus xy \oplus yz\,, \qquad (4)$$

henceforth called Logic II, which has the truth table shown in Table VI-12.

Theorems 1, 2, 5, and 7 of Section 2 are again applicable. Now, however, $2^n - 2$ is an upper bound on the sequence length obtainable, since there must be an even number of cycles (which rules out 2^n as a possible length), and neither 00...0 nor 11...1 will form a single-element cycle.

The symmetry between y and z, which was observed to hold for Logic I, is no longer present when Logic II is used. However, the

w	x	y	z	Period	Sequence
11	10	1	6	20	
11	7	4	10	15	
8	5	3	7	11	

Fig. VI-14. Three short cycles obtained using Logic II.

4. TABULATION OF EXPERIMENTAL DATA 147

operation of replacing x, y, and z by $n - x$, $n - y$, and $n - z$, respectively, still preserves cycle lengths.

As for Logic I, it is not difficult to specify the shortest cycle lengths obtainable from Logic II, where the *all 1's* vector has been used as the starting point. It is also possible to describe tap combinations yielding these shortest cycles.

If n is even, a cycle of length $(3n/2) - 1$ is obtained by letting $w = n$, $x = (n/2) + 1$, $y = (n/2) - 1$, and $z = n - 1$. If n is odd, a cycle of length $(3/2)(n - 1)$ is obtained by letting $w = n$, $x = (1/2)(n + 3)$, $y = (1/2)(n - 3)$, and $z = n - 1$. A "next-shortest" cycle for n odd, of length $2n-2$, is obtained by letting $w = n$, $x = n - 1$, $y = 1$, and $z = (1/2)(n + 1)$. These three situations are illustrated in Figure VI-14. Section 4.6 shows the set of periods obtained using Logic II with $n = 8$, and Section 4.7 shows the periods with $n = 11$. The starting vector in both cases was the *all 1's* vector. The data are presented in several forms (Tables VI-19 and VI-20, Figs. VI-19 and VI-20), as was done for Logic I.

4. TABULATION OF EXPERIMENTAL DATA

4.1. *Distribution of periods for Logic I and n = 11*

This section shows the periods obtained for $n = 11$, using Logic I, the initial state *all 1's*, and all possible tap combinations for x, y, and z. The periods are presented in numerical order in Table VI-13, and their distribution is portrayed in a histogram (Fig. VI-15).

Fig. VI-15. Distribution of periods into sixteenths of the interval from 1 to 2^{11} (Logic I, $n = 11$).

It's not surprising he couldn't do it. Because when one looks at nonlinear-feedback shift registers, one's effectively sampling the whole richness of the computational universe of possible simple programs. Back in the 1950s there were already theoretical results—mostly based on Turing's ideas of universal computation—about what programs could in principle do. But I don't think Sol or anyone else ever thought they would apply to the very simple—if nonlinear—functions in NFSRs.

And in the end it basically took until my work around 1981 for it to become clear just how complicated the behavior of even very simple programs could be. My all-time favorite example is rule 30—a cellular automaton in which the values of neighboring cells are combined using a function that can be represented as $p+q+r+qr$ mod 2 (or p XOR (q OR r)). And, amazingly, Sol looked at nonlinear-feedback shift registers that were based on incredibly similar functions—like, in 1959, $p+r+s+qr+qs+rs$ mod 2. Here's what Sol's function (which can be thought of as "rule 29070"), rule 30, and a couple of other similar rules look like in a shift register:

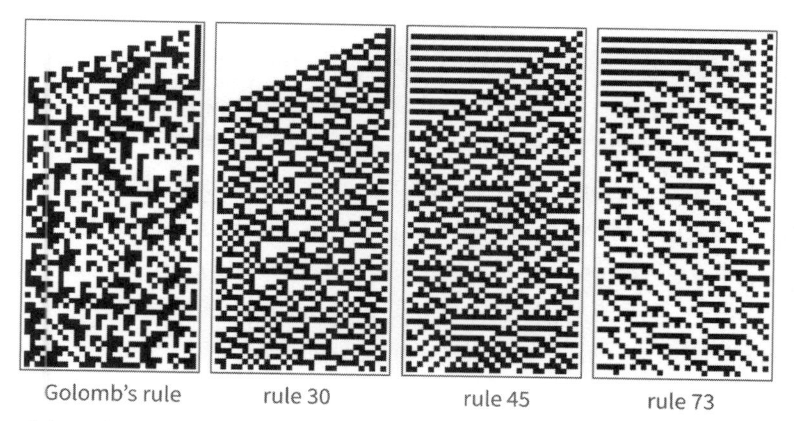

And here's what they look like as cellular automata, without being constrained to a fixed-size register:

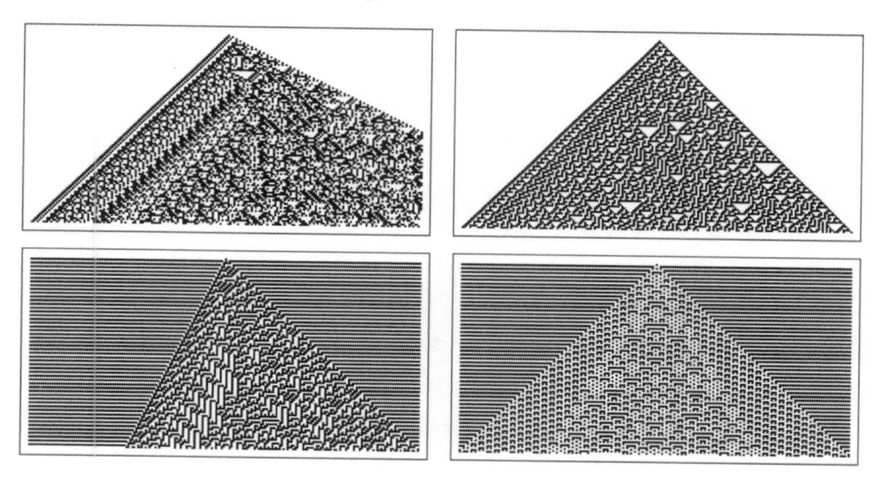

Of course, Sol never made pictures like this (and it would, realistically, have been almost impossible to do so in the 1950s). Instead, he concentrated on a kind of aggregate feature: the overall repetition period.

Sol wondered whether nonlinear-feedback shift registers might make good sources of randomness. From what we now know about cellular automata, it's clear they can. And for example the rule 30 cellular automaton is what we used to generate randomness for Mathematica for 25 years (though we recently retired it in favor of a more efficient rule that we found by searching trillions of possibilities).

Sol didn't talk about cryptography much—though I suspect he did quite a bit of government work on it. He did tell me though that in 1959 he'd found a "multi-dimensional correlation attack on nonlinear sequences", though he said that at the time he "carefully avoided stating that the application was to cryptanalysis". The fact is that cellular automata like rule 30 (and presumably also nonlinear-feedback shift registers) do seem to be good cryptosystems—though partly because of confusions about whether they're somehow equivalent to linear-feedback shift registers (they're not), they've never been used as much as they should.

Being a history enthusiast, I've tried over the past few decades to identify all precursors to my work on 1D cellular automata. 2D cellular automata had been studied a bit, but there was only quite theoretical work on the 1D case, together with a few specific investigations in the cryptography community (that I've never fully found out about). And in the end, of all the things I've seen, I think Sol Golomb's nonlinear-feedback shift registers were in a sense closest to what I actually ended up doing a quarter century later.

Polyominoes

Mention the name "Golomb" and some people will think of shift registers. But many more will think of polyominoes. Sol didn't invent polyominoes—though he did invent the name. But what he did was to make systematic what had appeared only in isolated puzzles before.

The main question Sol was interested in was how and when collections of polyominoes can be arranged to tile particular (finite or infinite) regions. Sometimes it's fairly obvious, but often it's very tricky to figure out. Sol published his first paper on polyominoes in 1954, but what really launched polyominoes into the public consciousness was Martin Gardner's 1957 Mathematical Games column on them in *Scientific American*. As Sol explained in the introduction to his 1964 book, the effect was that he acquired "a steady stream of correspondents from around the world and from every stratum of society—board chairmen of leading universities, residents of obscure monasteries, inmates of prominent penitentiaries..."

Game companies took notice too, and within months, for example, the "New Sensational Jinx Jigsaw Puzzle" had appeared—followed over the course of decades by a long sequence of other polyomino-based puzzles and games (no, the sinister bald guy doesn't look anything like Sol):

Sol was still publishing papers about polyominoes 50 years after he first discussed them. In 1961 he introduced general subdividable "rep-tiles", which it later became clear can make nested, fractal ("infin-tile"), patterns. But almost everything Sol did with polyominoes involved solving specific tiling problems with them.

For me, polyominoes are most interesting not for their specifics but for the examples they provide of more-general phenomena. One might have thought that given a few simple shapes it would be easy to decide whether they can tile the whole plane. But the example of polyominoes—with all the games and puzzles they support—makes it clear that it's not necessarily so easy. And in fact it was proved in the 1960s that in general it's a theoretically undecidable problem.

If one's only interested in a finite region, then in principle one can just enumerate all conceivable arrangements of the original shapes, and see whether any of them correspond to successful tilings. But if one's interested in the whole, infinite plane then one can't do this. Maybe one will find a tiling of size one million, but there's no guarantee how far the tiling can be extended.

It turns out it can be like running a Turing machine—or a cellular automaton. You start from a line of tiles. Then the question of whether there's an infinite tiling is equivalent to the question of whether there's a setup for some Turing machine that makes it never halt. And the point then is that if the Turing machine is universal (so that it can in effect be programmed to do any possible computation) then the halting problem for it can be undecidable, which means that the tiling problem is also undecidable.

Of course, whether a tiling problem is undecidable depends on the original set of shapes. And for me an important question is how complicated the shapes have to be so that they can encode universal computation, and yield an undecidable tiling problem. Sol Golomb knew the literature on this kind of question, but wasn't especially interested in it. But I start thinking about materials formed from polyominoes whose pattern of "crystallization" can in effect do an arbitrary computation, or occur at a "melting point" that seems "random" because its value is undecidable.

Complicated, carefully crafted sets of polyominoes are known that in effect support universal computation. But what's the simplest set—and is it simple enough that one might run across by accident? My guess is that—just like with other kinds of systems I've studied in the computational universe—the simplest set is in fact simple. But finding it is very difficult.

A considerably easier problem is to find polyominoes that successfully tile the plane, but can't do so periodically. Roger Penrose (of Penrose tiles fame) found an example in 1994. My book *A New Kind of Science* gave a slightly simpler example with 3 polyominoes:

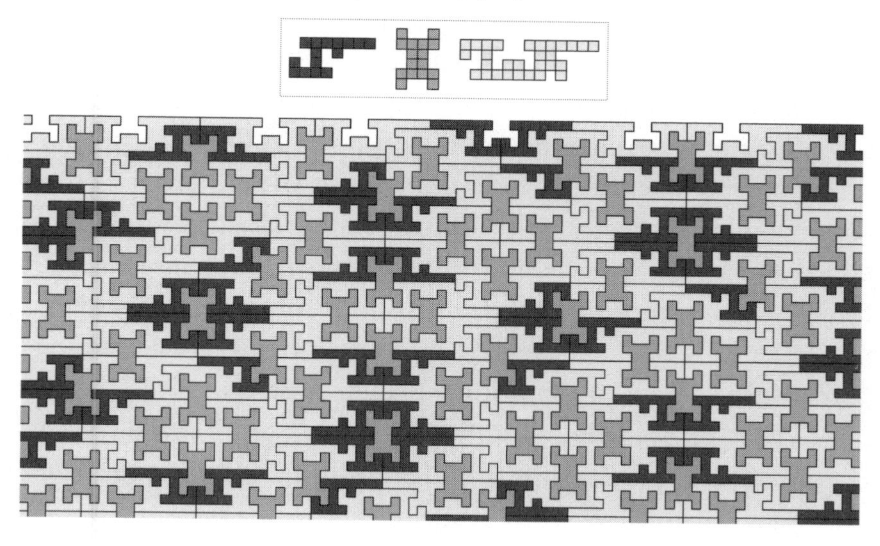

The Rest of the Story

By the time Sol was in his early thirties, he'd established his two most notable pursuits—shift registers and polyominoes—and he'd settled into life as a university professor. He was constantly active, though. He wrote what ended up being a couple of hundred papers, some extending his earlier work, some stimulated by questions people would ask him, and some written, it seems, for the sheer pleasure of figuring out interesting things about numbers, sequences, cryptosystems, or whatever.

Shift registers and polyominoes are both big subjects (they even each have their own category in the AMS classification of mathematical publication topics). Both have had a certain injection of energy in the past decade or two as modern computer experiments started to be done on them—and Sol collaborated with people doing these. But both fields still have many unanswered questions. Even for linear-feedback shift registers there are bigger Hadamard matrices to be found. And very little is known even now about nonlinear-feedback shift registers. Not to mention all the issues about nonperiodic and otherwise exotic polyomino tilings.

Sol was always interested in puzzles, both with math and with words. For a while he wrote a puzzle column for the *Los Angeles Times*—and for 32 years he wrote "Golomb's Gambits" for the Johns Hopkins alumni magazine. He participated in MegaIQ tests—earning himself a trip to the White House when he and its chief of staff happened to both score in the top five in the country.

He poured immense effort into his work at the university, not only teaching undergraduate courses and mentoring graduate students but also ascending the ranks of university administration (president of the faculty senate, vice provost for research, etc.)—and occasionally opining more generally about university governance (for example writing a paper entitled "Faculty Consulting: Should It Be Curtailed?"; answer: no, it's good for the university!) At USC, he was particularly involved in recruiting—and over his time at USC he helped it ascend from a school essentially unknown in electrical engineering to one that makes it onto lists of top programs.

And then there was consulting. He was meticulous at not disclosing what he did for government agencies, though at one point he did lament that some newly published work had been anticipated by a classified paper he had written 40 years earlier. In the late 1960s—frustrated that everyone but him seemed to be selling polyomino games—Sol started a company called Recreational Technology, Inc. It didn't go particularly well, but one side effect was that he got involved in business with Elwyn Berlekamp—a Berkeley professor and fellow enthusiast of coding theory and puzzles—whom he persuaded to start a company called Cyclotomics (in honor of cyclotomic polynomials of the form x^n-1) which was eventually sold to Kodak for a respectable sum. (Berlekamp also created an algorithmic trading system that he sold to Jim Simons and that became a starting point for Renaissance Technologies, now one of the world's largest hedge funds.)

More than 10,000 patents refer to Sol's work, but Sol himself got only one patent: on a cryptosystem based on quasigroups—and I don't think he ever did much to directly commercialize his work.

Sol was for many years involved with the Technion (Israel Institute of Technology) and quite devoted to Israel. He characterized himself as an "non-observant orthodox Jew"—but occasionally did things like teach a freshman seminar on the Book of Genesis, as well as working on decoding parts of the Dead Sea Scrolls.

Sol and his wife traveled extensively, but the center of Sol's world was definitely Los Angeles—his office at USC, and the house in which he and his wife lived for nearly 60 years. He had a circle of friends and students who relied on him for many things. And he had his family. His daughter Astrid remained a local personality, even being portrayed in fiction a few times—as a student in a play about Richard Feynman (she sat as a drawing model for him many times), and as a character in a novel by a friend of mine. Beatrice became an MD/PhD who's spent her career applying an almost mathematical level of precision to various kinds of medical reasoning and diagnosis (Gulf War illness, statin effects, hiccups, etc.)—even as she often quotes "Beatrice's Law", that "everything in biology is more complicated than you think, even taking into account Beatrice's Law". (I'm happy to have made at least one contribution to Beatrice's life: introducing her to her husband, now of 26 years, Terry Sejnowski, one of the founders of modern computational neuroscience.)

In the years I knew Sol, there was always a quiet energy to him. He seemed to be involved in lots of things, even if he often wasn't particularly forthcoming about the details. Occasionally I would talk to him about actual science and mathematics; usually he was more interested in telling stories (often very engaging ones) about personalities and organizations ("Can you believe that [in 1985] after not going to conferences for years,

Claude Shannon just showed up unannounced at the bar at the annual information theory conference?", "Do you know how much they had to pay the president of Caltech to get him to move to Saudi Arabia?", etc.)

In retrospect, I wish I'd done more to get Sol interested in some of the math questions brought up by my own work. I don't think I properly internalized the extent to which he liked cracking problems suggested by other people. And then there was the matter of computers. Despite all his contributions to the infrastructure of the computational world, Sol himself basically never seriously used computers. He took particular pride in his own mental calculation capabilities. And he didn't really use email until he was in his seventies, and never used a computer at home—though, yes, he did have a cellphone. (A typical email from him was short. I had mentioned last year that I was researching Ada Lovelace; he responded: "The story of Ada Lovelace as Babbage's programmer is so widespread that everyone seems to accept it as factual, but I've never seen original source material on this.")

Sol's daughters organized a party for his 80th birthday a few years ago, creating an invitation with characteristic mathematical features:

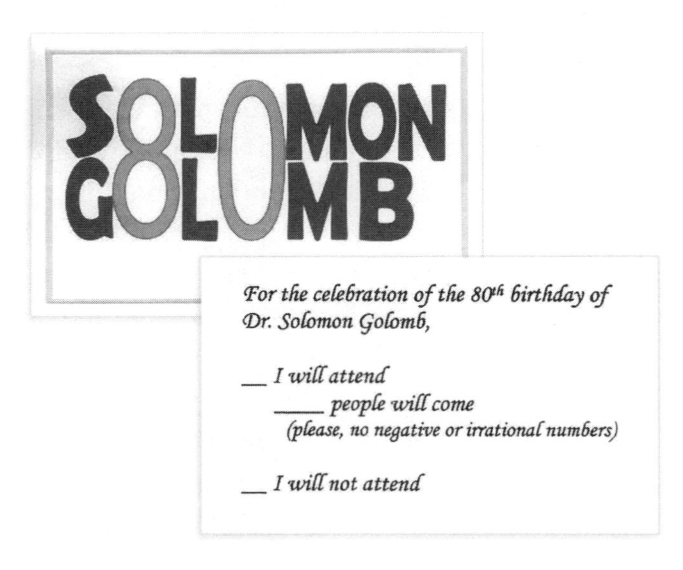

Sol had a few medical problems, though they didn't seem to be slowing him down much. His wife's health, though, was failing, and a few weeks ago her condition suddenly worsened. Sol still went to his office as usual on Friday, but on Saturday night, in his sleep, he died. His wife Bo died just two weeks later, two days before what would have been their 60th wedding anniversary.

Though Sol himself is gone, the work he did lives on—responsible for an octillion bits (and counting) across the digital world. Farewell, Sol. And on behalf of all of us, thanks for all those cleverly created bits.